NATIONAL DEFENSE RESEARCH INSTITUTE

Extending Depot Length and Intervals for DDG-51-Class Ships

Examining the 72-Month Operational Cycle

Roland J. Yardley, Daniel Tremblay, Brian Perkinson, Brenna Allen, Abraham Tidwell, Jerry M. Sollinger

Prepared for the United States Navy

For more information on this publication, visit www.rand.org/t/RR1235

Library of Congress Cataloging-in-Publication Data is available for this publication.
ISBN: 978-0-8330-9415-5

Published by the RAND Corporation, Santa Monica, Calif.
© Copyright 2016 RAND Corporation
RAND® is a registered trademark.

Cover image: *The guided missile destroyer USS Arleigh Burke (U.S. Navy photo by Journalist 2nd Class Patrick Reilly).*

Support RAND
Make a tax-deductible charitable contribution at
www.rand.org/giving/contribute

www.rand.org

Preface

Maintaining the fleet of surface combatants that the United States has built is challenging. The United States spends approximately $5 billion annually on ship depot maintenance. In addition to being costly, ensuring that the U.S. naval fleet operates at peak efficiency requires meticulous planning and execution of deployment schedules, crew training, and maintenance availabilities. The careful employment scheduling of ship deployments, short-term or long-term maintenance periods, and crew training ensures maximum readiness, efficiency, and expected service life of the ship. Combatant commanders' demand for surface combatant presence is high, and as the fleet has decreased in size, deployment lengths have increased to meet this demand.

Currently, the Navy is transitioning to a 36-month Optimized Fleet Response Plan operational cycle. However, with the constrained budget environment, the Navy is evaluating how best to manage training, maintenance, and presence needs to meet the deployment demands of these ships. The Director of Assessments (N81) within the Office of the Chief of Naval Operations (OPNAV) has asked RAND to examine the potential for extending the interval between depot maintenance periods, and for extending the length of time for the depot maintenance when the ship does undergo repairs and modernizations. This approach could achieve more deployed time for these ships, but it would be disruptive unless carefully planned before implementation. Many factors must be considered, from ensuring that crews receive the training needed before deployments, to ensuring that ships' service life is achieved, to meeting deployment demands, to balancing the tempo of operations for the crew. The Navy has also asked RAND to consider removal of the crew during the execution of maintenance to mitigate costs in this employment approach.

This research provides the Navy with an analysis of factors that must be considered in moving to a 72-month operational cycle that is followed by an extended maintenance period.

The research was sponsored by OPNAV N81 and conducted within the Acquisition and Technology Policy Center of the RAND National Defense Research Institute, a federally funded research and development center sponsored by the Office of the Secretary of Defense, the Joint Staff, the Unified Combatant Commands, the Department of the Navy, the Marine Corps, the defense agencies, and the defense Intelligence Community.

For more information on the RAND Acquisition and Technology Policy Center, see www.rand.org/nsrd/ndri/centers/atp or contact the director (contact information is provided on the web page).

Contents

Figures

Tables

Summary

Background and Purpose

As is the case with the other military services, the U.S. Navy expects to face a period of declining budgets and, indeed, is already experiencing them. Therefore, the Navy is seeking ways to operate its ships more cost-effectively. One approach might be to alter the deployment schedules of its surface vessels to get the greatest benefit in terms of operating efficiency and crew effectiveness. The Navy asked RAND's National Defense Research Institute to assess a 72-month deployment operational schedule followed by an extended depot maintenance period to determine what cost and efficiency benefits this approach might yield. The expectation is that a 72-month cycle will increase the time surface vessels are available to be deployed, allow for multiple deployments between depot availabilities, contain fewer basic training periods, and achieve some cost savings by removing the crews during the depot maintenance period. RAND researchers responded to the Navy's request by analyzing such a cycle using one class of ships, the DDG-51 *Arleigh Burke* class of destroyers.

Current Ship Cycles

The Navy has implemented three different ship cycles in recent years: 27, 32, and 36 months. Each cycle has the same major components: sustainment/deployment, maintenance, and training. The final component divides between basic and integrated or advanced training. The only thing that changes across the cycles is the amount of time devoted to each component. For example, in the 32-month cycle, 25 percent of the cycle's time is spent on deployment, and in the 36-month cycle, that fraction declines to 22 percent.

The 72-Month Operational Cycle

Under a 72-month cycle, a ship would go through a series of training and maintenance periods, deployments with carrier strike groups (CSGs), and unaccompanied deployments. At the end of its 72-month cycle, the ship would enter an extended maintenance period. During that extended maintenance period, most of the crew would leave the ship and marry up with the next ship coming out of the extended maintenance period. This cycle has the advantage of maintaining better cohesion among crew members, because they stay together longer and deploy together more often. It also offers a modest increase in training time and operational availability. The 36-month opreration cycle is shown in Figure S.1, and the 72-month operational cycle is shown in in Figure S.2.

Figure S.3 shows the differences in elements of the cycle that occur among the different deployment lengths. The yellow and green bars show two versions of the 72-month cycle, one with three deployments and one with four. The legend at the top of the figure shows the operational availability (A_o) percentages.

The 72-month operational cycles have a larger percentage of time devoted to integrated and advanced training, 18.6 and 22.6 percent, respectively, which is 3 and 7 percent more than with the 32- and 36-month cycles. They also show more deployment time, but by relatively small percentages. A_o is also greater, but again by relatively small percentages. The time spent in in sustainment (ready to deploy, but not deployed) is greatest in the current 36-month cycle.

Figure S.1
36-Month Operation Cycle for Surface Combatants

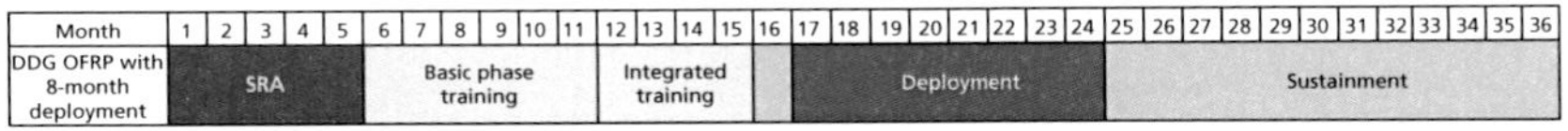
Month
1 2 3 4 5 6 7 8 9 10 11 12 13 14 15 16 17 18 19 20 21 22 23 24 25 26 27 28 29 30 31 32 33 34 35 36
DDG OFRP with 8-month deployment
SRA
Basic phase training
Integrated training
Deployment
Sustainment

NOTES: SRA = Selected Restricted Availability.

RAND *RR1235-S.1*

Figure S.2
Proposed 72-Month Operational Cycle for Surface Combatants

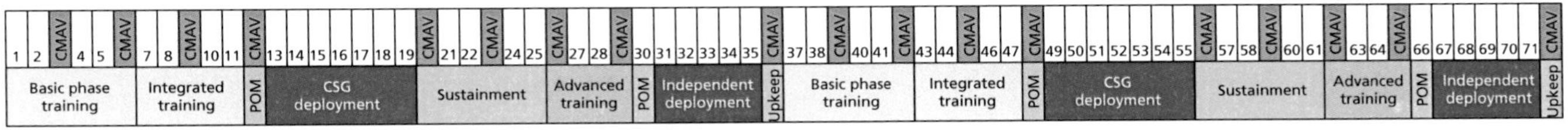
1 2 CMAV 4 5 CMAV 7 8 CMAV 10 11 CMAV 13 14 15 16 17 18 19 CMAV 21 22 CMAV 24 25 CMAV 27 28 CMAV 30 31 32 33 34 35 CMAV 37 38 CMAV 40 41 CMAV 43 44 CMAV 46 47 CMAV 49 50 51 52 53 54 55 CMAV 57 58 CMAV 60 61 CMAV 63 64 CMAV 66 67 68 69 70 71 CMAV
Basic phase training
Integrated training
POM
CSG deployment
Sustainment
Advanced training
POM
Independent deployment
Upkeep
Basic phase training
Integrated training
POM
CSG deployment
Sustainment
Advanced training
POM
Independent deployment
Upkeep

NOTES: CMAV = continuous maintenance availability; FRP = Fleet Response Plan; POM = pre–overseas movement.

RAND *RR1235-S.2*

Figure S.3
Fleet-Wide Time Spent in States for Differing Employment Cycles

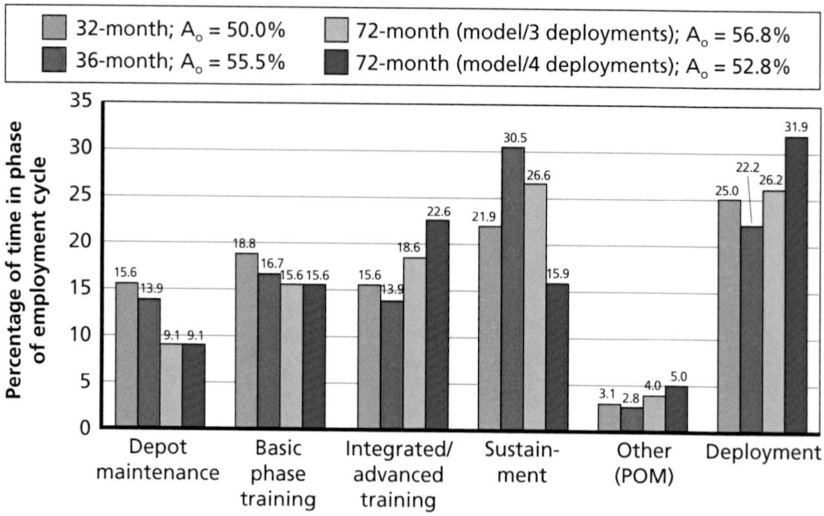

RAND RR1235-S.3

Crew Savings and Maintenance Costs

The primary motivation for this study was to explore the possibility of getting more deployments out of the DDG-51 fleet and to illuminate issues and challenges that the Navy must grapple with should it decide to proceed with this new operational cycle. A secondary motivation is to save costs. The potential cost savings would be the result of a reduction in the number of DDG-51 crews. As ships enter maintenance, they would be decrewed, and as long as there are ships in maintenance under the extended cycle, there will be more crews than ships. If the Navy were to divest itself of these surplus crews and reduce end strength to reflect this divestment, there would be a potential for cost savings. It is important to note that any reduction in crews will manifest in cost savings to the Navy and U.S. government only if end strength is reduced. Should the crews be divested but placed elsewhere in the Navy, no cost savings would be realized.

Savings from crew reductions is not the only component of cost that we considered. Maintenance demands also change under the new operational cycle, and we used our model output to capture the effects on cost. There are two aspects of maintenance costs to consider: the cost of additional continuous maintenance availability (CMAV) man-days under the extended cycle (because more continuous maintenance is needed with extended time between depot maintenance) and the change in cost of Chief of Naval Operations (CNO) maintenance availabilities over time as the fleet transitions to a 72-month operational cycle. We developed a model of the fleet today and, in the model, moved each ship to a 72-month operational cycle based on its current age and place in the 32-month cycle. We made several assumptions that we discussed with the sponsor that are key in modeling the maintenance and crewing in the transition of ships to the new cycle.

We highlight in Table S.1 the change in maintenance costs that would occur in the 72-month cycle. Ships begin to transition to the new cycle in FY 2016, and thus we begin to see the cost of additional CMAV man-days appear at this point, and grow year by year until FY 2022, when the last ship in the fleet transitions to the extended cycle. From FY 2022 on, the exact amount of additional CMAV man-days varies with the number of ships in depot availabilities. The change in CNO availability costs shows savings from FY 2018 through FY 2022, as ships in the new 72-month operational cycle bypass the depot availabilities they would have entered were they still in the 32-month cycle. Beginning in FY 2022, ships in the fleet begin to enter their first availability in the new cycle, and additional maintenance costs are incurred by FY 2023. The new maintenance package sizes are very large in size and duration, so the additional costs relative to the 32-month cycle are substantial.

We highlight in Table S.2 the combined effect of the change in maintenance costs and the potential crew savings that could be achieved in the 72-month cycle. The cumulative effect of these two maintenance components leads to a slight increase in costs during the first couple of years after ships begin to transition to the extended cycle, followed by four years of cost savings that result from ships bypassing the depot maintenance availabilities the fleet would normally undergo in the 32-month cycle, and finally an overall increase in cost because of

Table S.1
Changes in Maintenance Cost Under a 72-Month Operational Cycle, by Fiscal Year

	FY 2016	FY 2017	FY 2018	FY 2019	FY 2020	FY 2021	FY 2022	FY 2023	FY 2024	FY 2025	FY 2026
Cost of additional CMAVs in new cycle ($ millions)	>2.1	>5.5	>11.1	>16.7	>24.2	>30.6	>36.8	>40.0	>41.4	>43.1	>40.1
Change in CNO availability cost ($ millions)	0	0	<11.4	<32.1	<57.0	<86.7	<18.7	>130.6	>106.8	>67.5	>164.6
Total change in maintenance costs ($ millions)	>2.1	>5.5	<0.3	<15.4	<32.8	<56.1	>18.1	>170.6	>148.2	>110.6	>204.7

NOTE: Costs are in constant FY 2014 dollars.

Table S.2
Combined Maintenance Cost and Crew Reduction Savings Under a 72-Month Operational Cycle, by Fiscal Year

	FY 2016	FY 2017	FY 2018	FY 2019	FY 2020	FY 2021	FY 2022	FY 2023	FY 2024	FY 2025	FY 2026
Total change in maintenance costs ($ millions)	>2.1	>5.5	<0.3	<15.4	<32.8	<56.1	>18.1	>170.6	>148.2	>110.6	>204.7
Max annual total crew savings ($ millions)	0	0	0	0	0	0	<45–66	<45–66	<45–66	<45–66	<45–66

NOTES: Costs are in constant FY 2014 dollars. A range of costs are shown for crew savings and total (crew and availability) cost changes. This range represents potential annual crew savings to the Navy on the low end ($45 million per year) and savings to the government as a whole on the high end ($66 million per year). The difference accounts for benefits and entitlements that the government provides. The total change in costs reflects the combination of change in maintenance and crew savings.

the additional CMAV man-days and the increase and size and duration of maintenance availabilities in the 72-month cycle.

From FY 2023 through FY 2026, the increase in costs will be anywhere from $65.6 million to $159.7 million in a given fiscal year. However, this analysis was conducted to satisfy our primary motivation for a longer employment cycle and does not affect the results from changes to operational availability described earlier in this summary, nor does it affect any of the issues, challenges, and additional risks that the Navy would inherit should it choose to extend the operational cycle of the DDG-51 fleet. Our model does not project overall cost savings in the long run for the combined effect of savings as a result crew reductions and increases in maintenance costs. These costs should be weighed by marginal increases in the operational availability or deployments in the new cycle.

Results

Our analysis shows that the Navy can increase operational deployed time by shifting to a 72-month cycle. But, as mentioned above, the increase is not substantial. A 72-month cycle with three deployments increases deployed time by about 7 percent. A reduction in crew costs of $45 million to $66 million per year can occur, but not until FY 2022. However, these savings result from having fewer destroyer crews: Because of decrewing during maintenance, fewer crews are needed. This means that, to achieve the savings, the Navy must be willing to reduce its end strength. If it simply reassigns the crews to other personnel billets, no savings occur. Furthermore, under the 72-month cycle, maintenance costs climb. In part, this occurs because private providers provide more of the maintenance. Additionally, part of the crew would be needed to support the ship while it is in the extended maintenance period. Cost increases also occur because, under the extended cycle, additional CMAV man-days are required. In addition, maintenance costs change as the fleet transitions from the 32-month cycle to a 72-month operational cycle.

The Navy has been attempting to cope with a number of maintenance issues that have accrued over the years. One way it has attempted to deal with these is by varying deployment cycles. While some evidence suggests that improvement has occurred, the latest cycle change has not been implemented long enough to gauge its success. Nor is it clear that a 72-month cycle will resolve these issues.

Recommendations

Analysis of our results leads us to make the following recommendations, divided into two categories:

Maintenance Planning and Execution

1. Before going to a longer interval between depot maintenance, the Navy should correct impediments to availability execution.
2. Determine maintenance requirements. Senior Navy Engineering Duty authorities indicated that the Navy has not fully identified nor documented the conditions of surface combatants, particularly the condition of tanks. Tank maintenance is a major driver of maintenance and funding needs for depot work.
3. Develop a maintenance plan for the longer cycle. Navy maintenance authorities need to develop a plan that addresses the timing and sequence of maintenance in a longer operational cycle.
4. Increase continuous maintenance man-days; focus on life-cycle critical maintenance. With a longer interval between dedicated depot availability, increased continuous maintenance is needed to address both emergent maintenance demands and life-cycle critical maintenance.
5. Resource maintenance demands. A review of maintenance execution compared with the maintenance requirements contained in the DDG-51 technical foundation paper (TFP) (NAVSEA21, 2012c) indicates that ship's depot maintenance is funded below the requirement. The Navy should determine whether the TFP

requirement is actually the requirement and either fund it accordingly, adjust the requirement, or determine whether the risk (of not achieving expected service life) is acceptable to fund maintenance below the requirement.

6. Improve current maintenance planning and execution. Senior Navy maintenance experts indicated that current maintenance planning and execution are not as efficient and effective as they should be.

7. Evaluate the effect of maintenance demands on private providers. Little data are available that address the private supply of labor or the effect that a different maintenance cycle would have on the private providers of maintenance.

Training and Operations

1. If the Navy opts for a 72-month cycle, require ships to enter the cycle after CNO docking, and in a high state of material readiness. Senior maintenance authorities all voiced that ships must be in the highest state of material readiness to enter a cycle that requires a longer interval between depot maintenance periods. Moreover, a docking should precede this longer interval. DDG-51s are required to be docked at an eight-year interval. Exceeding that interval would raise the risk of catastrophic and costly failure of system components that must be maintained only in a docking availability.

2. Complete evaluations (dry-docking, tank conditions) of ship material readiness. Surface Maintenance Engineering Planning Program personnel indicated that an evaluation of just tank conditions would not be completed until the end of FY 2016, and the repair/maintenance of the tanks would be completed in FY 2022.

3. Award CNO availabilities in a fashion that allows for sufficient time for planning the work; the surface type commander must commit funding at the time of the award.

4. Fine-tune training to fit additional deployment needs. A new operational use of ships with an independent second and fourth

deployment in a 72-month cycle will increase training certification requirements for these additional deployments. The tailoring of training to meet the mission requirements of these additional deployments is needed.

5. Closely manage operating tempo. The Navy is exceeding tempo thresholds today with the current single eight-month deployment in the Optimized Fleet Response Plan (OFRP) cycle. Increased deployments in a 72-month operational cycle increases tempo, which requires close management of tempo thresholds and goals.

6. Use the model in this report to support analysis. The program developed can support fleet-wide analysis. The model and analysis that we have developed for the examination of the DDG-51 employment model can also be used for cruisers and amphibious ship. Moreover, the model we developed has the capability to provide a fleet-wide examination of maintenance and operational deployments, and how best to manage the various factors that are affected.

Acknowledgments

We would like to thank the staff of OPNAV N81 for their support, especially Carlton Hill, CDR Neil Sexton, and Stephen Williams. We also appreciate the guidance and insights provided by Chuck Werchado.

We thank RADM Dave Gale (PEO Ships) and RDML William Galinis (Commander, Navy Regional Maintenance Center) for their time and insights provided about surface ship maintenance.

We are thankful for our discussions with Erica Plath of OPNAV N96 on surface ship maintenance, Al Gonzalez of Commander, Fleet Forces Command, on operational tempo, Mike Harris of Commander, Surface Forces Atlantic, on surface ship maintenance scheduling, and Ray Weber of BAE Norfolk, who provided his thoughts from a surface maintenance provider perspective.

We also appreciate the insights provided by the staff of Commander, Naval Surface Forces, Pacific, including Ed White, Ted Serfass, Leon Stone, and Dan Rodgers.

We are indebted to our discussion with engineering specialists at the Surface Maintenance Engineering Planning Program (SURFMEPP), Norfolk, including CAPT Mike Malone, Tom Gallagher, and Fred Barnabei.

We appreciate the support provided throughout the project by RAND colleague Brad Martin. We are grateful to CAPT Andy Diefenbach (USN, Ret.) and RAND colleagues Jessie Riposo and John Schank for their thoughtful suggestions on an early draft of the report.

We also thank James Torr for editing the manuscript and Matthew Byrd for coordinating the document's production. We appreciate the administrative support provided by John Tuten.

The views expressed herein are our own and do not necessarily represent the policy of the Department of the Navy.

CHAPTER ONE

Introduction

Background, Purpose, and Audience

Surface combatants are the backbone of the U.S. Navy. These vessels enable the Navy to accomplish a variety of missions, whether supporting U.S. military operations around the world, protecting the world's commercial sea-lanes, engaging friendly nations, or providing humanitarian assistance. Ensuring that these vessels can accomplish their missions requires ships to operate on a carefully scheduled cycle that allocates time for training, maintenance, and modernization to achieve optimum readiness and desired forward presence.

Despite the Navy's crucial role in national security, it, too, has been subject to the budget cuts that have touched nearly every facet of the federal government. The tight budget that the Navy faces has prompted a review of the surface combatant employment cycle to ensure that these capital assets are being used efficiently and that their operational availability is being maximized in a cost-effective manner. Fiscal constraints on future Navy operations will likely lead to questions concerning the best employment model of surface combatants. An option to be considered is having fewer but longer depot maintenance periods. This approach would allow for a larger aggregated sustainment period where the ship is preserved and is capable of being deployed, support multiple deployments between depot maintenance, and allow for fewer basic training periods over the life of the ship. With this approach, the ship's crew (except for those required for safety, security, and maintenance oversight) would be removed during the extended depot period, and the entire ship would be turned over to the depot facility, similar to the aviation depot model.

The Navy asked RAND to examine the potential benefits of a 72-month operational cycle that increases the interval between depot maintenance and extends the duration of the depot availability. The ideal 72-month operational cycle would make more efficient use of ships by increasing the time surface vessels are deployed, allowing for multiple deployments between depot availabilities, and containing fewer basic training periods; secondarily, the ideal 72-month operational cycle would achieve some cost savings by the removal of crews during depot maintenance availability. The cost savings come from personnel reductions, because fewer crews would be needed. However, to achieve these savings, the Navy would have to cut its end strength. If the crews simply shift to somewhere else in the Navy, no savings occur.

The analysis in this report will interest a number of audiences. The stakeholders who would benefit from increased presence include the geographic combatant commanders; Commander, U.S. Fleet Forces Command (COMUSFLTFORCOM); Commander, U.S. Pacific Fleet (COMPACFLT); and the warfare enterprises with support from the lead technical authority, Commander, Naval Sea Systems Command (NAVSEA), which establishes the technical requirements. Regional maintenance centers (RMCs), ship maintenance activities, and detachments located in various major fleet concentration areas will also be interested, as will Commander, Naval Surface Forces, Pacific and Atlantic; the Navy's Afloat Training Groups; and private shipyards.

Approach

We examined a 72-month operational period followed by an extended depot availability for the DDG-51 *Arleigh Burke* class of destroyers.[1] We determined the expected loading (man-days) of the depot main-

---

[1] We began our research using a notional 18-month depot availability after the 72-month operational cycle. However, with the use of the DDG-51's technical foundation paper (TFP; NAVSEA21, 2012c), we calculated availabilities of varying length in the 72-month operational cycle, as discussed in Chapter Three.

tenance period to achieve planned expected service life (ESL), as well as the length, workload, and periodicity of a continuous maintenance availability (CMAV) program necessary to support a 72-month interval between depot availabilities (train and maintain once, deploy four times).[2]

To conduct a comprehensive study that considered what the challenges employing a 72-month operational cycle would entail in terms of maintenance, manpower, training, tempo of operations, and costs, we developed a study plan, conducted a literature review, determined the potential reduction in sea duty billets, calculated savings and costs, and identified costs, options, and the effects of removing the crew during depot maintenance. We consulted with multiple offices in the Navy that are responsible for surface ship training, manning, and operations, as well as industry experts who advised us on the private-sector role in maintenance availabilities. We also studied other sources of information, such as Visibility and Management of Operating and Support Costs, Center for Naval Analysis reports, the DDG-51 technical foundation paper (TFP; NAVSEA21, 2012c), and current Office of the Chief of Naval Operations (OPNAV) guidelines. We also built a fleet maintenance-scheduling model. This model makes it possible to transition and track all the ships in the class as they move out of the shorter cycles and into the 72-month cycle. It also tracks the time that a ship either is deployed or can be deployed (sustainment). And it presents the cost implications of moving to the 72-month cycle, specifically noting whether any cost savings might accrue. We describe this model in Chapter Six, detailing its design and underlying assumptions.

---

[2] Continuous maintenance is a process that involves the near continuous flow of work candidates to the most appropriate maintenance level and maintenance activity for accomplishment. A vital part of continuous maintenance is the scheduling and accomplishment of work outside of Chief of Naval Operations (CNO) availabilities. This allows the ship to be consistently maintained at acceptable readiness levels. Work performed during a CMAV includes inspections, condition-based upkeep, and minor repairs. The work takes approximately three weeks to complete and is scheduled once every three months.

This final report addresses the issues listed below:

- the maintenance requirements that would need to be addressed in the longer cycle
- the alterations in crewing policies needed to make this cycle succeed
- changes to training schedules to accommodate a 72-month cycle
- deployment schedules that ensure the surface combatants can accompany the carrier strike group (CSG) and independent deployments and not exceed personnel tempo (PERSTEMPO) and operating tempo (OPTEMPO) thresholds
- the potential cost savings that could result from the adoption of this employment cycle.

We conducted qualitative interviews with senior leaders and subject-matter experts who are knowledgeable of maintenance, training, manpower, and ship scheduling and operations. The interviews consisted of open-ended discussions with regard to the challenges and opportunities in moving ships to a 72-month operational period followed by an extended depot period, during which the crew is removed from the vessel (and joins a vessel just emerging from depot maintenance). A list of those we interviewed appears in Appendix A, along with a list of the pertinent Navy references we drew on.

Assumptions

To perform this analysis, we made several assumptions about maintenance, crewing, training, and employment of DDG-51 *Arleigh Burke*–class ships that were informed by subject-matter experts, Navy references, and discussions with our sponsor. We also used the study team's experience and judgment in making these assumptions. Table 1.1 outlines the key assumptions that we made to frame the analysis and the justification for each.

We assume that ships would enter the 72-month cycle after a CNO docking period, since ships must be in excellent material condition upon entering the extended cycle. The docking period would provide the needed maintenance. The reliance on CMAVs with increased

Table 1.1
Key Research Assumptions and Justification in the Analytic Approach to Evaluating a 72-Month Operational Period for DDG-51 *Arleigh Burke*–Class Destroyers

Key Assumption	Justification
Maintenance	
DDG-51 TFP provides authoritative depot maintenance requirements.	DDG-51 TFP, February 2012
Ships enter a 72-month operational period after a CNO docking availability.	SURFMEPP and RMC interview
Increased reliance on CMAVs; CMAV maintenance man-days doubled to equate roughly to FDNF CMAV man-days; CMAVs focus on life-cycle critical maintenance.	Commander, Naval Surface Force Pacific, maintenance officer; RMC; SURFMEPP interviews
Private maintenance providers can support new depot maintenance and CMAV approach.	Study team assertion
CNO availabilities greater than 6 months in duration must be bid coast-wide.	TFP, and discussion with SURFLANT scheduler
Ship will move to maintenance facility for duration of depot availability.	Discussion with project monitor.
Every extended depot maintenance period will be a docking in the new cycle.	RMC, NAVSEA21, SURFMEPP interview
Deferred maintenance will contain a "fester factor."	SURFMEPP
The work normally done by the crew during a depot will be performed by the depot maintenance provider. The number of depot maintenance man-days will be increased by the amount of man-days of effort normally assigned to the ship's force (ship will be decrewed during the depot period).	Project description, research, and study team assumption
Level of effort expended in extended depot maintenance (after the 72-month cycle) is based on the average number of man-days of all docking avails in the DDG-51 TFP.	Study team assumption
Depot maintenance man-days will be reduced by LCC work done during CMAVs in operational period.	Study team assertion
Fester factor = 6% per annum	VADM Burke: "$2 Billion Backlog in Surface Ship Maintenance Hard to Dig Out Of"[a]

Table 1.1—Continued

Key Assumption	Justification
Crewing	
During the depot availability at the end of the 72-month operational period, crew will be removed, except for those needed for the safety and security of the ship and for maintenance management of the availability. Notional number of crew members = ~50 personnel.	Project description requirement
Crews will not change homeports. Crews on ships going into maintenance will man a ship coming out of maintenance in our modeling. There will be a 1-month gap in changing ships for the oncoming crew.	Discussion with project monitor at Interim Project Review meeting
Training	
Unit-level training must be done every 36 months.	COMUSFLTFORCOM N1; personnel rotation
Ships conduct integrated training with the CSG before the 1st and 3rd deployments and advanced training before the 2nd and 4th deployments.	Training requirements in the Surface Force Readiness Manual and OFRP instruction
Employment	
1st and 3rd deployment will be 7-month duration with a CSG to maintain alignment; 2nd and 4th deployment are independent deployments of ~ 4.5 months in duration.	Project description, CFFC N1 assertion of Navy's desire for 7-month deployment length, and alignment with CSG.
Nondeployed steaming days = 24 days a quarter.	FY 2015 Navy budget and Overseas Contingency Operations request
PERSTEMPO/OPTEMPO limits—the deployment length for the 2nd and 4th deployments will be adjusted to keep the ship within OPTEMPO and PERSTEMPO limits.	Study team assertion

NOTES: SURFMEPP = Surface Maintenance Engineering Planning Program; FDNF = Forward Deployed Naval Force; SURFLANT = Surface Force Atlantic; NAVSEA21 = Naval Sea Systems Command, Deputy Commander, Surface Warfare.

[a] "Burke: $2 Billion Backlog in Surface Ship Maintenance Hard to Dig Out Of," InsideDefense.com, March 22, 2013.

focus on lifecycle critical maintenance performed during these periods will be required to enable a longer period between dedicated maintenance availabilities. We also take as a given that private yard maintenance providers can support the new depot maintenance approach, as well as increased CMAV support. CNO availabilities that are greater than six months in duration, by law, must be bid coast-wide, and the ship will move to the maintenance facility for the duration of the maintenance period. Every maintenance period in the new cycle will be a docking period, so no dry-docking availabilities will be bypassed. Deferred maintenance will cost more to fix in the future than the present, and we call this a "fester factor." For our calculations, we use a 6 percent per annum fester factor.[3] By the same token, maintenance done early will be done at a discount; however, a degradation factor for moving maintenance forward increases the maintenance costs, because more maintenance is now required. After planned maintenance, degradation occurs to the system reliability, availability, or benefit between neighboring planned maintenance cycles.

The work done by the crew during a maintenance period, called the ship's force work list (SFWL), will be transferred to the maintenance provider for completion, because the crew is taken off the ship during the maintenance availability. We assume the level of effort expended in extended maintenance periods to be equal to the average number of man-days that is performed in all docking periods, as described in the DDG-51 TFP. Finally, the total number of depot maintenance man-days will be reduced by the life-cycle critical maintenance done during CMAVs performed in the operational period.

Our crewing assumptions are that, during the depot availability at the end of the 72-month operational period, the crew will be removed, except for those who are required to maintain the safety and security of ship. The number of crew members needed to maintain the safety and security of the ship is approximately 50 people. We also

---

[3] While a "fester factor" in ship maintenance is widely acknowledged to exist, its precise value is debatable. We have used comments from VADM William Burke from March 21, 2013: "If you let it go, it's festering. It festers at about 6 percent. That which costs you $100 today, in a year it's going to cost you $106. We've got some empirical data on this" ("Burke: $2 Billion Backlog in Surface Ship Maintenance Hard to Dig Out Of," 2013).

assume that the crew will not change homeport. The entire crew will rotate to a ship coming out of maintenance. There will be a one-month gap to allow time for the crew to rotate from the ship going into maintenance to manning the ship coming out of maintenance.

For crew training, we assume that unit-level training (ULT) must be done every 36 months. Ships will conduct integrated training with the CSG before the first and third deployments. Crews will perform advanced training, tailored to their deployment needs, before the second and fourth deployments.

For the ship's employment, the first and third deployments will be seven months in duration and be performed with the CSG. The second and fourth deployments in a 72-month cycle are independent deployments. The number of nondeployed steaming days per quarter used for OPTEMPO calculations is 24.[4] To maintain the ship and crew within OPTEMPO and PERSTEMPO limits, we adjusted the time deployed for the second and fourth independent deployments accordingly.

Organization of This Report

Chapter Two examines the background and evolution of the employment scheduling of surface ships. Chapter Three discusses past and present maintenance challenges and addresses a potential approach to meet maintenance demands for surface combatants maintenance in a 72-month cycle. Chapter Four covers manpower assigned to *Arleigh Burke*–class destroyers and associated costs, as well as training entitlements that must be met to prepare these ships and crews for deployed operations. Chapter Five explores the effect of a 72-month cycle on OPTEMPO versus Navy guidelines and thresholds. Chapter Six describes the model that was used in examining the effect of a 72-month cycle on the *Arleigh Burke*–class destroyer fleet. Chapter Seven presents our findings, conclusions, and recommendations. Appendix A provides a listing of personnel whom we interviewed, and Appendix B provides background on surface combatant maintenance.

---

[4] The FY 2015 Department of the Navy budget allocates 20 days per quarter for nondeployed ships, which is supplemented with an Overseas Contingency Operations request for an additional four days per quarter (Department of the Navy, 2014).

CHAPTER TWO

DDG-51 Fleet and Employment Scheduling

This chapter describes four ship employment cycles: 27, 32, 36, and 72 months. For each cycle, we discusses the time allocated to deployment, maintenance, and training. We also provide the rationale for why the Navy might want to go to a 72-month cycle.

The Challenge

Altering the schedule of surface combatants is complex, and key ship employment factors must be taken into account. Time must be allocated for training, maintenance, and modernization, and these must be balanced against operational needs. Sacrificing crew training, maintenance, and modernization for increased deployments imposes opportunity costs in terms of the crew's effectiveness, the material condition of the vessels, and the ability of both the crew and the ship to perform to expected standards.

The Navy has modified the employment schedule for surface vessels in recent years. Ships transitioned from a 27-month cycle to a 32-month cycle in 2006 and are currently transitioning to a 36-month cycle. Expanding the employment cycle still further to 72 months in an effort to increase operational deployments requires careful consideration and scheduling of crew training and maintenance and modernization needs.[1]

---

[1] The 27-, 32-, and 36-month cycles that are referenced include depot maintenance periods as well as operational periods. However, the 72-month cycle that we will discuss in this report

The current Navy approach to the Optimized Fleet Response Plan (OFRP) is to first perform maintenance and unit-level and integrated training, and then deploy for eight months in the 36-month cycle (Gortney, 2014). With one deployment per cycle, a longer cycle reduces the number of deployments over a ship's service life but extends the length of each deployment. Figure 2.1 compares the number of deployments that can be made over the ESL of ships in different cycle lengths. The ESL of DDG-51 *Arleigh Burke*–class destroyers Flight I and II is 35 years, and Flight IIA has an ESL of 40 years. Figure 2.1 illustrates that, with one deployment per cycle, a Flight IIA destroyer will deploy up to 18 times over its ESL in the 27-month cycle, while under a 36-month cycle it will deploy approximately 13 times. The point is that there are trade-offs to extending the length of operational cycles (such as OFRP) while maintaining a single deployment per cycle. A key trade-off, as indicated in Figure 2.1, is fewer deployments in the longer cycle over the ESL of the ship.

Figure 2.1
Notional Number of Deployments (one deployment per cycle) for DDG-51-Class Ships, by Cycle Length and Ship Type

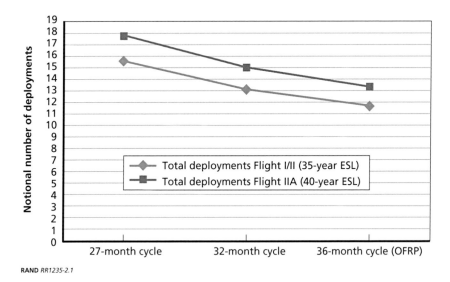

RAND RR1235-2.1

is only an operational cycle. A depot maintenance period of varying length will follow the 72-month operational cycle.

In longer cycles, the deployment length (for single deployment per cycle) must increase to match the total deployed time over a ship's ESL. For example, as illustrated in Figure 2.2, an eight-month OFRP deployment in a 36-month cycle provides the same amount of deployed time as a six-month deployment in 27-month cycle. The arrow indicates the comparison of months deployed in 27-month cycle with a six-month deployment, compared with an eight-month deployment in a 36-month OFRP cycle. With longer employment cycles and the demand for surface combatant presence remaining steady, ships must remain deployed longer to equal the time for which they are deployed under shorter cycles with shorter deployments.

Using single eight-month deployments in the 36-month OFRP cycle yields 106.7 months of total deployed time over the ESL of the ship.

Our discussions with fleet authorities indicate that the Navy is trying to reduce deployment lengths to seven months. As Figure 2.2 illustrates, shorter deployment lengths reduce the total number of months deployed for a ship as the cycle length increases.

Figure 2.2
DDG-51 Flight IIA Total Months Deployed over Service Life, Single Deployment, in 27-, 32-, 36-Month Cycles

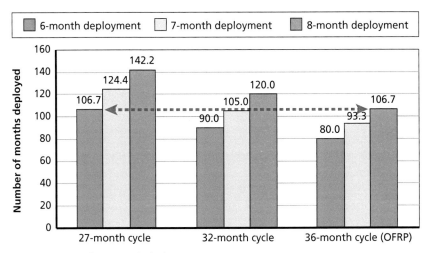

SOURCE: Research team calculations.

RAND RR1235-2.2

Using seven-month deployments in a 32-month cycle over the life of a ship is roughly equivalent to using eight-month deployments in a 36-month cycle. Using six-month deployments in a 27-month cycle achieves the same deployed time over the ship's ESL as using eight-month deployments in a 36-month cycle, and using eight-month deployments in a 27-month cycle yields the most total deployed months of all the options shown in the figure. Going forward, the question to be addressed is, "Can the Navy achieve more operational deployments with a new approach, and what are the costs, challenges, and risks?"

DDG-51 *Arleigh Burke*–Class Destroyers

The DDG-51 *Arleigh Burke*–class destroyer is a multimission ship. Currently the class has 62 ships, with more under construction. Below, we briefly describe the ship's capabilities and characteristics.

The DDG-51 *Arleigh Burke*–class guided missile destroyers provide a wide range of warfighting capabilities in multithreat air, surface, and subsurface environments. These ships respond to Low Intensity Conflict/Coastal and Littoral Offshore Warfare (LIC/CALOW) scenarios as well as open-ocean conflict independently or as units of CSGs, expeditionary strike groups, and missile defense action groups. Named after famed World War II officer and former Chief of Naval Operations Arleigh Burke, DDG-51-class ships provide outstanding combat capability and survivability characteristics while considering procurement and life-cycle support costs (NAVSEA, 2015).

The *Arleigh Burke* class is currently composed of three flights of ships: Flight I (DDG-51 through DDG-71), II (DDG-72 through DDG-78), and IIA (DDG-79 and above). Flight IIAs are slightly longer than the Flight I and II, and they can embark helicopters. The armament of the flights is as follows (NAVSEA, no date-a):

Flights I and II (DDG-51–78)
Standard Missile (SM-2MR)
Vertical Launch ASROC (VLA) Missiles
Tomahawk

Six MK-46 Torpedoes (from two triple tube mounts)
Close In Weapon System (CIWS)
5″ MK 45 Gun
Evolved Sea Sparrow Missile (ESSM)

Flight IIA (DDG-79+)
Two LAMPS MK III MH-60 B/R Helicopters with Penguin/
    Hellfire Missiles
MK 46/MK 50 Torpedoes

Ship Employment Schedules

The employment cycle of a ship is a month-by-month schedule of major types of conditions or employment under which a ship operates. Broadly speaking, the major employment categories that a ship normally passes through are sustainment, deployment, maintenance, and training (either basic or integrated or advanced). Sustainment generally means that a ship is ready to deploy, but has not been tasked to do so. Over the past several years, surface combatants have changed the length of their employment cycles. For example, the *Surface Force Readiness Manual* (COMNAVSURFPACINST/COMNAVSURFLANTINST 3502.3, 2012a) outlines the Fleet Response Training Plan 27-month cycle. This cycle is used in the Navy's readiness manual and indicates that maintenance is part of the continuum of training, and not necessarily a stopping and starting point. Figure 2.3 illustrates the elements of this cycle, with the sustainment/deployment period running from months 12–22 and training from months 1–11.

Ships normally follow this cycle of maintenance, shakedown (a period of time for material assessment, watch team training, and certification following maintenance), basic training, integrated training (combining unit warfare skills into a single CSG in a multiwarfare environment) or advanced training (mission-specific training for ships not assigned to a strike group), and deployment and sustainment.

In 2006, the Navy moved to a 32-month cycle, illustrated in Figure 2.2. This cycle follows the similar cycle of maintenance and

Figure 2.3
27-Month Cycle for Surface Combatants

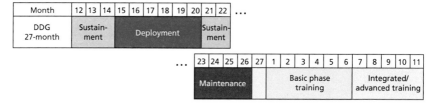

RAND *RR1235-2.3*

training (basic and integrated/advanced) followed by a deployment. However, some of the details differ. The Navy no longer schedules a shakedown period, and it has added a pre–overseas movement period (POM) during which crew can take leave and prepare for overseas movement. The 32-month cycle includes a longer sustainment period than the 27-month cycle. Figure 2.4 shows a five-month sustainment time, with sustainment periods both before and after a deployment.

The maintenance periods in the 32-month cycle are described in the TFP for DDG-51-class ships (NAVSEA21, 2012c). These availabilities will serve as the basis for required and prescribed maintenance of a ship through its ESL as we explore a 72-month operational cycle. These availabilities are as follows:

- Selected Restricted Availability (SRA): An SRA is a maintenance period during which selected modernizations are also executed. SRAs are nominally 12–13 weeks in duration, depending on the

Figure 2.4
32-Month Cycle for Surface Combatants

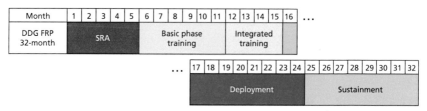

RAND *RR1235-2.4*

length of the operational schedule (NAVSEA21, 2012c, p. 22). While 12–13 weeks is the nominal time, our discussions with Navy fleet schedulers indicate that SRA maintenance availabilities are currently scheduled for 16 weeks in duration, and more than 50 percent of ships exceed this time allotted.

- Docking Selected Restricted Availability (DSRA): A DSRA is an SRA that requires dry-docking to perform certain maintenance and modernization tasks. They are 14–18 weeks in duration, depending on the length of the operational schedule (NAVSEA21, 2012c, p. 49).

- Depot Modernization Period (DMP): A DMP is an important availability focused on upgrading high-priority warfare systems. DMPs typically occur about halfway through a ship's ESL and can last for over a year (NAVSEA21, 2012c, p. 60).

- CMAV: According to the DDG-51's TFP, CMAVS "are intended for accomplishment of inspections (assessments), upkeep (condition-based), and minor repairs (including emergent)." CMAVs have an approximate duration of three weeks and are scheduled once per quarter within a fiscal year (NAVSEA21, 2012c, p. 7).

Current 36-Month Employment Cycle

The Navy adopted a 36-month cycle in 2013, known as OFRP. This plan, which is illustrated in Figure 2.5, is meant to create a more agile and flexible fleet that can surge on short notice while still meeting global force management (GFM) commitments and preserving the sustainability of forces in the long run (OPNAVINST 3000.15A, 2014).

OFRP also enhances fleet readiness by aligning the surface combatants employment schedule with that of the CSG (OPNAVINST 3000.15A, 2014). OFRP contains four phases: maintenance (also known as an SRA); basic and integrated, or advanced training; deployment; and sustainment. This operational framework creates a readiness cycle that aligns forces across the fleet both operationally and administratively. OFRP extends the 32-month cycle by four months, and the 27-month cycle by nine months. Over the course of these 36 months,

Figure 2.5
36-Month Current Employment Cycle—the Optimized Fleet Response Plan

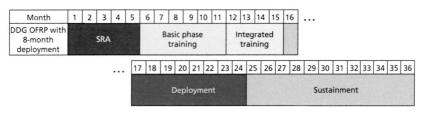

RAND *RR1235-2.5*

22.2 percent of the cycle is spent on deployments, which are eight months in duration; 13.9 percent of the cycle is spent in crew training; and another 13.9 percent is spent conducting maintenance and modernization. Compared with the shorter 32-month cycle, during which 25.0 percent of the cycle is spent on deployment, the OFRP reduces the percentage time deployed.

Table 2.1 depicts the major differences between cycles. The most substantial difference is the amount of sustainment time provided by the different cycles.

Table 2.1
Comparison of Time Spent in Major Employment Events, by Cycle Length

	Cycle Length		
	27 Months	32 Months	36 Months
Sustainment	5	8	12
Deployment	6	8	8
Maintenance	4	5	5
Training	11	10	10
Other/POM	1	1	1

NOTES: Training includes both basic and advanced. Numbers are rounded.

Reevaluating the Current Employment Cycle

Two motives drive consideration of a new cycle. The first is meeting the national security demands the Navy faces. Increasing operational availability is a top priority for the Navy, and so the Navy must operate the surface combatant fleet more efficiently. The second motivation is the need to maximize the use of these capital assets. The fiscal constraints under which the Navy currently operates will likely continue for the foreseeable future. Drastic budget-cutting measures, such as sequestration, make it challenging for the Navy to decide how best to reduce spending. The Navy is considering a 72-month operational cycle, which would support increased deployments over a 72-month operational period, employ a long depot period at the end of the operational period, and remove the crew during depot, potentially allowing the Navy to operate the current fleet of surface combatants with a smaller number of crews.

Overview of the 72-Month Operational Cycle

RAND was asked to analyze a 72-month operational period followed by an extended maintenance period. The cycle we designed is illustrated in Figure 2.6.

This 72-month operational cycle would increase the forward presence of DDG-51 *Arleigh Burke*–class ships by deploying them four times over a 72-month period. The new cycle allows a crew to train once and deploy twice in the first 36 months of the cycle, before repeating basic phase training starting in month 37. While one of the study's original objectives was to design a cycle that would allow crews to deploy four times between ULTs, we found this approach to be untenable.[2] When a vessel enters a depot availability at the end of the new cycle, the crew departs and then boards a vessel coming out of a depot availability, at which time both the crew and the ship restart the

---

[2] Discussions with Commander, Fleet Forces Command (CFFC), manpower authorities indicate that a high turnover rate will occur after 36 months, and thus a need for ULT.

Figure 2.6
72-Month Operational Cycle

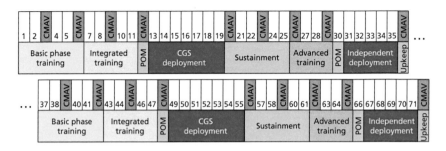

RAND *RR1235-2.6*

72-month cycle. The first and third deployments are modeled to last for seven months, with the second and fourth lasting 4.5 months. A caretaker crew would remain onboard to maintain the safety and security of the ship and maintenance oversight, and its composition and tasking are discussed in a following chapter.

CHAPTER THREE

Addressing Maintenance Demands in the 72-Month Cycle

The purpose of this chapter is to address a potential approach to surface combatant maintenance in a 72-month operational cycle.[1] Maintenance and modernization needs must be considered when altering operational cycles for surface combatants. These ships are maintenance-intensive assets. Performing maintenance and modernizations can be costly and time-consuming, and thus reduce the time a vessel can be deployed. In practice, limited maintenance budgets often lead to dealing with only the most critical issues. As more maintenance is neglected (or deferred), the effects of deferred maintenance continually increase and can cut short the effective lifespan of the system and result in a loss of the expected economic value of the asset (Malone et al., 2014). Neglecting maintenance and modernization detracts from a ship's readiness and performance, increases the likelihood that a vessel will not reach its ESL, and ultimately drives up repair costs, because maintenance is bypassed. The appropriate balance must be struck between meeting immediate, mission-essential maintenance needs and ensuring that life-cycle repairs are not neglected because of short-term operational needs or budget constraints.

---

[1] For a discussion on surface combatant maintenance, including its current status, please refer to Appendix B.

Modeling Ship Depot Maintenance in the 72-Month Cycle

The 2012 TFP for DDG-51 *Arleigh Burke*–class ships captures the life-cycle requirements that define the baseline per ship and identifies the maintenance requirement to reach ESL. The DDG-51 TFP

- is based on a long-range maintenance schedule that was developed in conjunction with the Class Maintenance Plan and historical standard cost trends based on return cost data. The long-range maintenance schedule does not include an aging factor or ship alteration cost data.
- provides a solid, defensible foundation of technical requirements, CNO and continuous maintenance man-day estimates, and recommended CNO availability durations.
- is an in-depth technical analysis that identifies maintenance periodicity requirements in terms of the ship's life cycle.
- provides a foundation for integrating modernization into availability planning.
- provides a framework for identifying, scheduling, and tracking maintenance requirements.
- serves as a management tool for long-range planning, budgeting, and resource allocation.
- does not address organizational level (ship's force) maintenance tasks (NAVSEA21, 2012c).

Drawing from the TFP for the DDG-51 class and from official Navy guidelines concerning maintenance practices, such as OPNAVINST 4700.7L (2010), we determined which Navy maintenance policies must be taken into account when developing a 72-month operational cycle. One such policy is that CNO-scheduled private-sector depot-level maintenance availabilities of six months or longer must be bid coast-wide (OPNAVINST 4700.7L, 2010). Because depot availabilities following the 72-month operational schedule will be a minimum of seven months (as will be shown in the discussion of this study's model outputs), all depot availabilities for the DDG-51 class must be bid coast-wide at the end of the new cycle. Thus, potential homeport shifts may occur if the winning maintenance provider is not physically located in the ship's homeport.

Another policy RAND considered is one that defines the amount of labor in man-days available, stating, "The available sustainable labor force in each assigned CONUS [continental United States] homeport . . . is 650 man-days/day (13,000 man-days/month) for a DDG-51 Class ship" (OPNAVINST 4700.7L, 2010). This amount of available labor is based on a five-day workweek. Furthermore, Navy policy stipulates that CMAVs are to be conducted once every FY quarter for an approximate duration of three weeks. As will be discussed later in this chapter, the 72-month operational cycle RAND proposes fulfills this Navy policy and actually enhances the role of CMAVs in the overall approach to conducting maintenance.

Number of DDG-51s in CNO Maintenance Availabilities
The TFP outlines a DDG-51 32-month maintenance plan, and this plan can be modeled to provide a baseline to examine potential changes. Figure 3.1 provides a baseline model of the DDG-51 fleet that is in service (not in a maintenance availability), in maintenance availability, and, of those in maintenance availability, the number that are in a docking availability. These data are based on maintenance

Figure 3.1
DDG-51s in Service, Maintenance, and Docking Availabilities, and Out of Maintenance, per Technical Foundation Paper

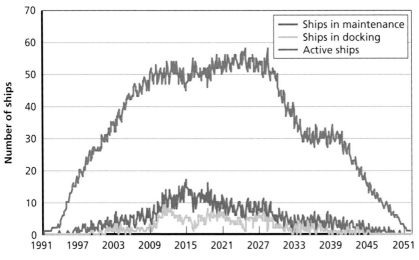

plans in the DDG-51 TFP for a 32-month cycle that consists of availabilities described in Chapter Two. This baseline model can be used to identify potential changes to the operational cycle and the effect on the number of ships in maintenance. As illustrated, approximately ten ships, or one-sixth of the DDG-61 fleet, are in maintenance at a time, and roughly one-half of those ships in maintenance are in a docking availability. These data provide an opportunity to understand the estimated number of ships fleet-wide that are in maintenance and the number that are operational.

Developing Maintenance Availabilities in a 72-Month Operational Cycle

The time line in Figure 3.2 illustrates when life-cycle maintenance is conducted in a 32-month cycle. Drawing from the TFP for the DDG-51 class and official Navy guidelines concerning maintenance practices (OPNAVINST 4700.7L, 2010), this chart outlines the major constraints we have identified in terms of what manpower is available

Figure 3.2
Life-Cycle Maintenance Schedule for DDG-51 Flight IIA on a 32-Month Cycle

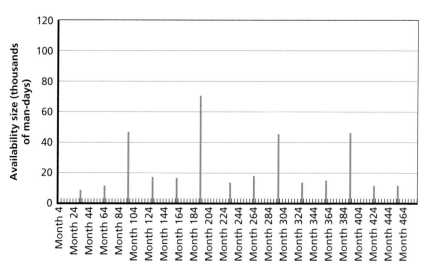

for maintenance, how long key maintenance packages require (nominally) for completion, and the nominal amount of expected continuous maintenance scheduled into a given fiscal year. Critical for discussions concerning actual Navy capacity for executing the lengthy (12-month) maintenance availabilities is the fact that availabilities of six months or more may be bid coast-wide. As mentioned above and as we will show below, this is a useful provision, because it opens up a much larger pool of piers and dry-docking facilities to any given DDG-51.

Timing and Size of CNO Depot Maintenance Availabilities in a 72-Month Cycle

The TFP is the current authority for life-cycle maintenance plans for the fleet. The underlying assumptions on which the 72-month schedule is based are that life-cycle maintenance for the class, which is laid out in the TFP, will not change under a new operational cycle, because no work disappears and no "new" work is generated. In short, the maintenance requirements laid out in the TFP will be the same under the new 72-month operational cycle. Also, phasing of availabilities is altered to fit the new operational/maintenance cycle, though rephasing and delaying work introduces the fester factor and degradation factor for specific availabilities. The life-cycle maintenance schedule also assumes a greater reliance on CMAVs to perform maintenance between availabilities, which now occur every six years after the previous docking availability has been completed. Furthermore, ships will be decrewed during depot availabilities, meaning the maintenance provider must perform the SFWL, which is work normally assigned to the crew while the ship is in a maintenance period.

The 72-month operational cycle is currently followed by a docking availability (as will be discussed later, the length of availabilities in the new cycle will vary over the course of a ship's ESL). Because all new availabilities are dockings, the first four availabilities in the new 72-month cycle are associated with the four docking availabilities from the 32-month cycle. The actual duration of the availability will be determined by the size of the new maintenance package. The factors that will determine the size of the new availability package are the SRA man-days in the operational period, the docking availability man-days in the operational period, the fester factor/degradation factor, and the SFWL man-days, minus the additional CMAV man-days conducted in the operational period. Figure 3.3 illustrates the factors that are considered when determining the size of the new maintenance package.

Figure 3.4 illustrates how the new packages will be spaced over the life cycle of ships on the new cycle compared with the 32-month cycle. As can be seen by comparing the number of man-days spent on availabilities for the 32-month cycle with those for the 72-month cycle, the availabilities for the 72-month cycle are significantly larger. These larger availabilities account for the fester factor (6 percent) and the deg-

Figure 3.3
Factors Determining Depot Maintenance Man-Days

SOURCES: NAVEAS21, 2012c; OPNAVINST 4700.7L, 2010.
RAND RR1235-3.3

Figure 3.4
Comparison of 32-Month Cycle and 72-Month Cycle Depot Package Sizes

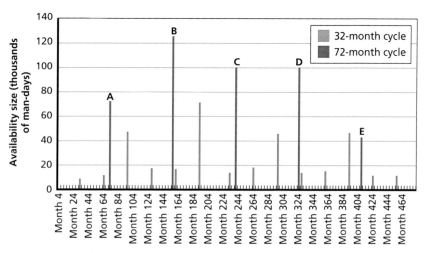

SOURCES: NAVEA21, 2012c; OPNAVINST 4700.7L, 2010.
RAND RR1235-3.4

radation factor (6 percent), and they combine the shorter dockings and pier-side maintenance periods built into the 32-month cycle.

We computed the sizes of the availabilities based on the data from Figure 3.4. For example, we calculated availability B in the 72-month operational cycle according to the equation in Figure 3.3 as follows: The second docking availability in the 72-month cycle will correspond to the second docking availability in the 32-month cycle. According to Figure 3.3, the first term contains the man-days for SRA 2-1 and 2-2, which are 17,400 and 17,100 man-days, respectively. Next we include the man-days from the DMP in the 32-month cycle, which is the second docking. The DMP contains 70,700 man-days, according to the TFP. The third term will contain a fester factor for performing SRA 2-1 late and a degradation factor for performing the first docking availability, DSRA-1, early—a total of 8,343 man-days. The fourth term, representing SFWL, which is transferred to the yard, totals 22,800 man-days for the duration of this new availability. Finally, we deduct 10,872 CMAV man-days that have taken place during the operational phase. The final result is an availability that will take 125,471 man-days.

We computed the durations by using the average number of man-days for docking availabilities, as provided in the DDG-51 TFP (NAVSEA21, 2012c), and applying those man-days to the computed size of the computed maintenance packages. Table 3.1 provides the size and duration of CNO maintenance availabilities in the 72-month cycle, keyed to the A through E periods in Figure 3.4.

Navy maintenance authorities indicated that newly commissioned ships could start a 72-month operational cycle after a post-shakedown availability, as ships will be in a high state of material condition (being new) and most maintenance and modernization needs have been met.

Table 3.1
Computed Size and Duration of CNO Maintenance Availabilities Under a
72-Month Cycle for DDG-51 *Arleigh Burke*–Class Destroyers

	A	B	C	D	E
Size (thousands of man-days)	72	125	100	101	43
Duration (months)	7	13	10	10	4

Older ships can also transition to the new cycle from the 36-month cycle. As mentioned, Navy authorities argued that other ships in the fleet should be in excellent material condition before transitioning to the 72-month operational cycle. Therefore, the ideal time for ships to transition is after completing a docking availability.

Surface combatants will transition to the new cycle at different times depending on their age, as shown in Figure 3.5. The DDG-51 fleet is composed of ships of different ages, and we configured transition points for various ships based on their ages. There are four transition points to the new 72-month cycle, based on the age of a ship. The blue vertical lines represent the number of man-days under current maintenance plan, and the red vertical lines represents the number of man-days under the 72-month operational cycle. Figure 3.5 illustrates when DDG-51s would begin a 72-month cycle. For example, a ship that is 196 months (16.3 years) in age would transition to a 72-month cycle after a scheduled CNO docking availability under the 32-month cycle. We used this approach and these transition points to model the entire DDG-51 fleet.

Figure 3.5
Transition Points of DDG-51 *Arleigh Burke*–Class Ships to a 72-Month Cycle

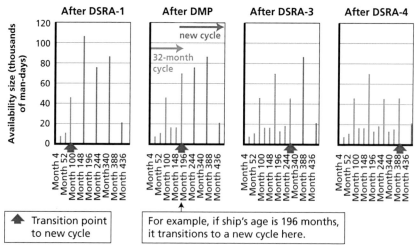

SOURCE: NAVSEA21, 2012c; OPNAVINST 4700.7L, 2010.

As depicted in Figure 3.5, ships enter the 72-month operational cycle after a DSRA or after a DMP. Once on the 72-month cycle, depot availabilities begin to occur as the new cycle dictates. For some ships, such as those nearing the end of their life cycles, adopting the 72-month operational schedule may not be feasible or worthwhile. Undergoing an expensive overhaul during the last years of a ship's life cycle simply so it can transition to a 72-month operational cycle, especially if the ship has less than 72-months remaining before being decommissioned, may be an unwise use of Navy resources. Whether ships nearing the end of their life cycle should transition to the 72-month operational cycle or undergo significant depot availability can be decided based on the necessity of the final availability.

Increasing the Focus of CMAVs to Life-Cycle Critical Maintenance

In addition to reforming how depot availabilities are planned and executed, we recommend a different approach to conducting CMAVs. As indicated above, CMAVs are scheduled availabilities for surface force ships, normally two to six weeks in duration and normally scheduled once per nondeployment quarter when the ship is in port (OPNAVINST 4700.7L, 2010). Interviews with SURFMEPP indicate that CMAVs typically last three weeks. CMAVs compete with training and operational demands, and as a result CMAV man-days are often deferred, which decreases material readiness and increases the size of the maintenance and modernization package that must be executed during the next CMAV or depot availability. This situation is referred to as a deferred maintenance backlog. Typically, backlogs are difficult to eliminate, because work that is deferred to a later maintenance period supplants work that should be conducted at that time; catching up on deferred maintenance while executing current maintenance becomes increasingly difficult. The backlog is exacerbated by a fester factor, which is the increase in the severity of a needed repair that is deferred. The longer a maintenance need goes unaddressed, the more costly and time-consuming it is to fix, hence worsening the maintenance backlog.

As previously mentioned, our proposed 72-month operational cycle recommends increasing the number of maintenance man-days required for CMAVs for the surface combatants to match the number of man-days allocated for forward deployed naval forces (FDNF).[2] This will double the number of man-days used to execute maintenance in CMAVs

Summary

In this chapter, we relied on the DDG-51 TFP for the 32-month cycle to describe what maintenance both Flight I/II and IIA ships need to achieve their ESL. Modifying these maintenance demands to fit into a 72-month operational cycles shows the need for fewer, though much longer, docking availabilities. In aggregate, these availabilities actually require a larger number of total man-days across the life of a ship, due to deferred maintenance contributing to a fester or degradation factor, as well as transfer of the SFWL away from the crew to the yard conducting the availability. This increase in the total number of man-days will drive an increase in the cost for the total maintenance necessary for a ship to reach ESL.

Now that we have established the increase in maintenance cost, we must evaluate this against the potential benefits: There will be cost savings should the Navy be able to reduce the total number of crews necessary to operate the fleet in the future under this new operational cycle, as well as the potential for additional operational availability and deployments under the new cycle. Chapter Six will describe the modeling approach that we used to clarify these issues. Only with the output of the modeling will a full cost-benefit analysis of extending the operational cycle of the fleet be possible.

---

[2] FDNF ships get nearly double the continuous maintenance due to the pace of operations. In a 72-month cycle, with potentially three or four deployments, DDG-51-class ships will operate very similarly to FDNF ships.

CHAPTER FOUR

DDG-51 Manpower and Training

This chapter addresses the manpower and training needed to meet the employment demand of DDG-51-class ships under a 72-month rotation. We first discuss the number of people involved in crewing and their associated costs. We then describe the process of moving crews off ships and how many would have to remain with the ship when it goes into its extended maintenance period and the drawbacks associated with that process. We then describe what performing the maintenance would entail. We conclude the chapter with a discussion of the training of the crews. We provide our hypothetical 72-month training cycle and explain some of the rationale for that cycle, and we discuss what training exceptions might be employed. Finally, we discuss some of the policies associated with transferring crews from one ship to another.

Manpower Demands

Ships are nothing without the crew, and the new cycle requires personnel adjustments. The reforms needed will at times conflict with ingrained Navy culture but will ultimately allow the Navy to increase operational availability and realize reduced crewing costs after implementing the 72-month operational cycle. One such personnel reform is crew swapping, which the Navy specifically asked to be included in RAND's proposed cycle. The 72-month operational cycle includes decrewing a vessel at the end of the cycle and then placing the crew onto a vessel coming out of depot availability. The cycle we developed allows for a one-month gap at the beginning and end of depot avail-abilities to give the crew time to decrew and then recrew a ship.

Another assumption of this study, which aligns with Navy policy, is that the crew coming off a ship entering depot availability will not have to change homeport. Homeport shifts are disruptive to crews and their families and cost additional money to fund. On the other hand, depot availabilities in the new cycle will be longer than six months, and therefore maintenance work must be bid coast-wide. To keep crews in their homeport, ships coming out of depot will move to where their crews are located rather than the crew changing homeport to meet the ships. Figure 4.1 shows the crew complements of the DDG-51 series of ships by their officer and enlisted components. The Flight IIA ships, for example, have a slightly smaller crew than the earlier flights (one officer and nine enlisted personnel).

DDG-51 Manpower Costs

The Office of the Secretary of Defense uses the Full Cost of Manpower (FCoM) tool, which is designed to generate cost estimates associated

Figure 4.1
DDG-51 Manning Levels, by Flight

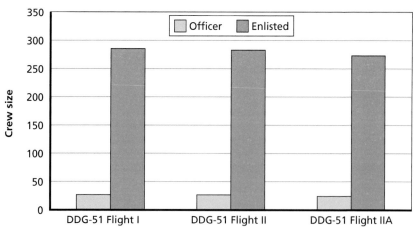

SOURCE: Navy Manpower Analysis Center (NAVMAC), DDG-51 Ship Manpower Documents.
RAND RR1235-4.1

with U.S. Department of Defense (DoD) manpower—military, civilian, and contractor personnel. The FCoM tool shows estimated costs for the DoD component, DoD, and the federal government. FCoM provides a consistent approach for all DoD employees to estimate the fully burdened costs of manpower.[1] The major DoD cost driver is health care, and the major federal government cost drivers are military retirement contributions and U.S. Department of Veterans Affairs benefits.

Designed to reduce effort needed to estimate costs associated with DoD manpower, the FCoM tool relies on user input to determine specific attributes associated with military, civilian, and contractor personnel, such as occupation/specialty, rank/grade, length of service, and location. The tool automatically estimates the total annual cost for each type of manpower submitted by the user.

Using the tool and DDG-51 manning documents that were provided to us by the Navy Manpower Analysis Center, we entered the grades of the DDG-51 crew, by flight, into the tool to derive the manning costs. Figure 4.2 provides the DDG-51 annual manpower costs for the Navy, combined Navy and DoD, and total federal manpower costs, by flight.

As is clear in this graph, each crew is expensive, costing the Navy approximately $30 million annually. Despite the potential cost savings of reducing the number of crews needed to operate the surface fleet, the Navy must plan for and make adjustments for future end-strength reductions to realize savings from such an approach.

Our analysis indicates that the cost savings that might accrue from reducing the number of crews would be $44 million annually for the Navy and $66 million annually for the federal government, which absorbs some costs for Navy personnel. How these costs were arrived at is explained in Chapter Six.

Removing Crew During Maintenance

Private contractors will execute most of the maintenance package in the new cycle. As a result, the majority of the crew will not need to stay with the ship in depot; however, some Navy personnel will need

---

[1] Derived from FCoM website (not available to general public).

Figure 4.2
DDG-51 Annual Manpower Costs, by Flight

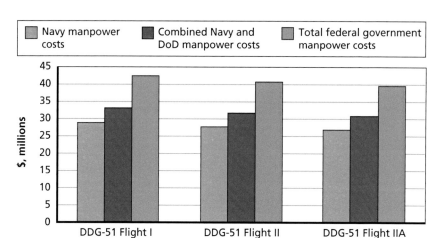

SOURCE: Office of the Secretary of Defense FCoM tool.
NOTE: Costs are in constant FY 2014 dollars.
RAND RR1235-4.2

to remain behind with the ship for safety, security, and maintenance management. Senior naval officers and previous research suggest that approximately 50 (or one-sixth of crew) personnel are needed to conduct these tasks while a vessel is undergoing a depot availability.[2] To ensure that as much crew cohesiveness is preserved during the decrewing process as possible, a permanent detachment should be based on the depot facility to execute the tasks just described. This will ensure that most of the crew coming off a ship entering a depot availability will stay together as they board a ship coming out of a maintenance period.

The depot maintenance detachment is a necessary component of maintenance execution. These personnel are tasked with coordinating maintenance support with the shipyard and maintaining safety and security on the ship by meeting an antiterrorism and force protection requirement, fire and flooding response, sounding and security, and security/quarterdeck watch standing. Furthermore, personnel will be

---

[2] Derived through an interview and materials provide with OPNAV N96 personnel.

required to operate and maintain mechanical, electrical, and ventilation systems on board the ship and fulfill administrative, supply, and food preparation roles.

The decrewing approach in the new cycle is not completely foreign to the Navy. A 2006 Center for Naval Analyses study examined the effect of partially decrewing surface ships undergoing maintenance availabilities of at least seven months. In this study, a skeleton crew of petty officers and junior offices stayed with the ship in depot to conduct vital repairs and fire watch (Choi et al., 2006). Cost savings from this approach were found to be $295.9 million for FY 2008. The initial cost savings were $448.4 million, but $152.5 million was then spent to contract out the SFWL. The study also looked at readiness issues that stem from partial decrewing during long maintenance availabilities and indicated that records from the 1980s show that readiness problems do occur, but that they are usually overcome before deployment.

Navy Manpower Required When Ships Go into Depot Maintenance

Because ships will be decrewed as they enter depot availability, contract maintenance providers will execute the SFWL. The size of the work package that those providers will have to supply can be computed by multiplying the available workforce by the Navy's standard workweek by the time available to conduct the maintenance. We estimate that the ship's E-5s and below (approximately 200 personnel) are available to conduct maintenance and that approximately four hours per day are spent working. This means that 800 man-hours can be accomplished per day (or 100 man-days per day), and 2,000 man-days can be completed per month (100 man-days x 5 days per week x 4 weeks per month). The private yard provider of maintenance at increased cost will execute the maintenance man-days allocated to the crew.

However, the contractor workforce does not accomplish all the tasks that must be done. A maintenance crew of Navy personnel is needed to maintain the safety and security of the ship. As noted, we estimate the DDG-51 maintenance crew size to be approximately 50 personnel. The missions and tasking that the crew will perform during the maintenance availability include the following:

- coordinate maintenance support with shipyard
- maintain safety and security of the ship
- perform anti-terrorism/force protection requirements
- provide fire/flooding response
- stand sounding and security watches
- security and quarterdeck watch-standing duties
- operate and maintain mechanical, electrical, and ventilation systems
- maintain (in lay-up) systems as needed
- perform administrative and supply tasks.

Figure 4.3 breaks down the number of personnel in the estimated maintenance crew by grade. We posit that the maintenance crew could be composed of shore personnel permanently stationed in the region where maintenance is conducted. Naval reservists could support this mission.

These personnel can come from the crew of the ship, or they can come from a depot detachment, which would have to be created to supply them. In either case, they must factor into any calculations of manpower savings.

Potential Drawbacks to Decrewing

The Navy has experimented with different types of personnel rotations in the past. One rotation that has been tested multiple times is sea swapping, which allows ships to remain deployed longer by swapping out the crews mid-deployment, thus allowing the ship to stay in theater while a new crew is flown in to replace the previous crew.[3] Despite the savings potential of these crew rotations, the value of these policies is debated within the Navy. A 2007 Congressional Budget Office report provided an overview of the different approaches to crew swapping the Navy has considered, including a 2002 Navy experiment with sea swapping for ships at the end of their ESL in an effort to increase for-

---

[3] For example, the personnel rotation plan for the Navy's new Littoral Combat Ship is to rotate crews (currently from San Diego to a forward location). The Navy will use three crews for two ships (one at the forward location, one at the CONUS homeport), with the goal of reducing crew fatigue and maximizing the use of the forward-deployed ship.

Figure 4.3
Number of DDG-51 Maintenance Crew Members, by Grade

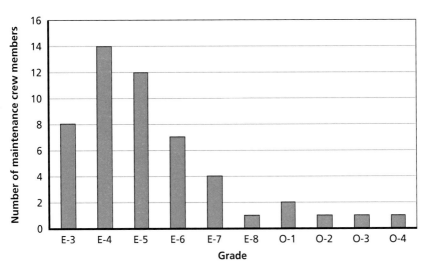

ward presence (Congressional Budget Office, 2007). One of the disadvantages of this experiment was that crews had less familiarity with the vessels on which they served and consequently did not appropriately maintain the ships. A later study found that forward presence increased 40 percent when sea swapping was used for ships on 18-month deployments (meaning crews were swapped every six months for 18 months rather than the ship returning to base every six months to get a new crew) (Congressional Budget Office, 2007). Unfortunately, this crew rotation increased the number of shore-based personnel needed, thus detracting from the savings achieved.

Another disadvantage was noted in a 2005 Center for Naval Analyses study of a Naval Surface Forces Pacific (SURFPAC) experiment in sea swapping (Choi, Birchler, and Duquette, 2005). This study suggested that morale suffers as a result of this crew rotation because sailors are typically transported into or out of theater by plane, thus denying the crew the opportunity to visit interesting ports during a deployment. The study also claimed that crew workloads increase near the end of deployment, as the crew readies the ship to be handed over to the oncoming one, thus creating another quality-of-life issue. If the Navy were to have to expend resources to ensure retention in the face of these issues, it would cut into the potential savings these personnel policies are meant to deliver.

The 2005 Center for Naval Analyses study also pointed out another reality of achieving cost savings with crew swapping, which is that, in many cases, the cost savings come from having fewer crews. In SURFPAC's experiment with sea swapping, most of the savings came from the Military Personnel, Navy (MPN), account (Choi, Birchler, and Duquette, 2005). SURFPAC cited these savings because the experiment assumed Navy personnel end strength was reduced to reflect the need for fewer crews.

In short, the savings from crew rotations occur only if the crew is released from the Navy, with a resulting reduction in end strength. The 72-month operational cycle would enable the Navy to operate its existing surface fleet with fewer crews, but the Navy would have to carry out these reductions to reap the rewards.

Training Requirements

A previous section of this chapter described the manpower considerations that would arise from a decision to move to a 72-month cycle. This section discusses the training considerations. The Navy has a scripted set of training requirements to move from training individuals, to crews, to ships operating in conjunction with other ships. These training events must occur at specific intervals, be synchronized with deployment type, and account for the normal turnover that occurs as a result of such things as enlistment terms and tour lengths. Whatever cycle length the Navy settles on must take these considerations into account.

Basic Phase Training and Integrated/Advanced Training

OPNAVINST 3000.15A (2014) lays out the purpose of basic phase training and integrated/advanced training. According to the document,

The basic phase focuses on development of unit core capabilities and skills through the completion of basic-level inspections, certifications, assessments and visits of personnel, equipment, supply, and ordnance readiness. Units and staffs that have completed the basic phase are ready for more complex integrated or advanced training events, or appropriate tasking.

Activities to be completed during integrated training are also established:

The purpose of integrated phase training is to synthesize individual units and staffs into aggregated, coordinated strike groups (or other combined-arms forces) in a challenging multi-dimensional threat, realistic warfare environment. . . . Upon completing of the integrated phase, strike groups and other combined arms forces will be certified to deploy. (OPNAVINST 3000.15A, 2014)

Advanced training is tailored training to prepare ships for independent deployments.

Navy Policies Define Training Time Entitlements

Training is crucial if Navy personnel are to achieve their missions. As a result, the Navy has laid out important training guidelines to ensure that sailors and crews are prepared to face the missions and the challenges they will encounter while deployed. Despite the importance of training, allocating time for training in an operational cycle takes away from time that can be spent on deployment. Training is conducted before a ship begins a deployment. Although training occurs throughout deployment, crews are trained and must be certified by training authorities before they depart homeport for an extended deployment. Training is a necessary component in a ship's employment cycle, and Navy policies provide training time entitlements for unit-level, integrated, and advanced training.

The Navy has issued guidelines (COMPACFLT/COMUSFLT-FORCOMINST 3501.3D, 2012) that Pacific Fleet uses for scheduling the appropriate amount of time to accomplish necessary milestones during a given operations cycle. While these are not hard requirements, they are generally accepted guidelines that provide a set of scheduling constraints, which we used in developing the 72-month schedule. These guidelines are as follows:

1. Basic training for surface combatants in all operational configurations requires a nominal **24 weeks**.
2. Ships without a dedicated CNO availability between scheduled deployments remain in sustainment and will execute a basic training certification validation to support certification extension.
3. Integrated training for CSG or ballistic missile defense deployments requires a nominal **21 weeks**.
4. Advanced training for independent deployments requires **16 weeks** of training.

The goal of integrated training is to combine individual unit and staff warfare skill sets into a single cohesive strike group capable of operating within a challenging, multiwarfare, joint, multinational, and interagency environment and to train staffs to take a tactical leadership role.

Advanced training applies to independent deployers that are not part of a CSG. The goal of advanced training is to conduct advanced core and mission-specific training to meet combatant or Navy component commander requirements.

Training in the 72-Month Operational Schedule

Based on the Navy's guidelines and the OPNAV Director of Assessments' (N81's) request that RAND develop a schedule allowing four deployments between basic phase training periods, also known as ULTs, we made several assumptions about how training should be conducted in a 72-month operational schedule. The first is that conducting ULT once and deploying four times is untenable. Our discussions with senior CFFC personnel representatives indicate that nearly all the crew that start an operational cycle on a ship rotate off within 36 months, and approximately one-third of the crew leaves each year.[4] In particular, all officers rotate off a ship within 36 months. These rotations are the result of regular sea-shore rotations, end of active obligated service, and other losses that reduce crew continuity. As a result, we concluded that ULT must be done every 36 months; otherwise, it is unlikely that the majority of the crew members on board after that point will have trained together.[5] As seen in Figure 4.4, basic phase training is conducted before the first and third deployments (yellow blocks).

As is also shown in Figure 4.4, integrated training with the CSG (light green blocks) is conducted after basic phase training and before the first and third deployments, which are the deployments made with the CSG. Integrated training synthesizes individual units and staffs into well-functioning strike groups in a realistic warfare environment (OPNAVINST 3000.15A, 2014). Required inspections, certifications, assessments, visits, and training are also completed during this phase, and necessary levels of personnel, equipment, supplies, and ordinance readiness are also achieved. Strike groups and other combined arms forces are certified to deploy after completing integrated training, as

Figure 4.4
RAND-Constructed Notional 72-Month Operational Cycle

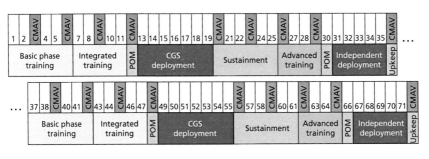

RAND RR1235-4.4

---

[4] Based on discussion with CFFC N-1.

[5] Based on discussion with CFFC N-1 regarding personnel rotation in the fleet.

they will have demonstrated their ability to operate in joint and coalition operations.

Advanced training (salmon-colored blocks) takes place prior to the second and fourth deployments, which are independent deployments. The purpose of advanced training is to execute core and mission-specific training (OPNAVINST 3000.15A, 2014). Proficiency in mission areas must be demonstrated, and the crew will be certified to deploy upon completion of this phase.

Basic Phase Training Objectives and Extension

Based on guidelines established for Surface Forces, Pacific, from COMPACTFLT/COMSURFLTFORCOMINST 3501.3D (2012, Chapter 3: Fleet Training Response Plan, Section 2(b): Basic Phase), there appear to be no preapproved scenarios for when a unit may skip basic training. That being said, completion of the basic phase is, according to the guidelines, based on showing proficiency in the following tasks (COMNAVSURFPACINST/COMNAVSURFLANTINST 3500.11, no date):

1. Operate and maneuver safely.
2. Operate and communicate with other similar type units as well as cross-platform using installed systems.
3. Defend own unit.
4. Restore and ensure survivability and sustainability of unit capabilities.
5. Effectively employ own unit equipment, weapons, and sensors.
6. Employ and demonstrate unit-level tactics, techniques, and procedures in individual warfighting/mission area(s) utilizing the most appropriate mix of live, virtual, and constructive training methods in accordance with established type commander training and readiness policy.

One of the primary goals of this strategy is for each ship to have a standard, predictable training path throughout the Fleet Training Response Plan. This predictability is necessary to synchronize the various maintenance, training, and operational requirements. Circumstances may require deviations from the hypothetical Fleet Training Response Plan cycle. Although a range of possibilities exists, three primary variations are described below:

1. **Full Basic Phase.** Ships that conduct a scheduled CNO availability and have sufficient training time available before the next deployment will execute the established training and certification plan.
2. **Abbreviated Basic Phase.** When a ship is not allotted sufficient time to complete a full basic phase following a CNO availability, a tailored training plan will be established based on the results of a readiness evaluation (READ-E). This tailored plan will provide training in those mission areas assessed by the type commander as below minimum acceptable standards.
3. **Certification Extension.** Ships without a dedicated CNO availability between scheduled deployments remain in the sustainment phase and will execute a certification validation to support certification extension. The certification validation will be a comprehensive assessment of all assigned warfare area certifications. Mission areas validated below certification criteria will receive additional training in order to maintain certification (COMNAVSURFPACINST/COMNAVSURFLANTINST 3502.3, 2012, pp. 4–18).

The nominal weeks allocated (24–25 weeks based on function of DDG-51 unit in question) are not explicitly built in to the guidelines as a "hard and fast" amount of time required—thus, in theory, a unit does not have to use all of the allocated weeks, provided it meets the aforementioned milestones. Cutting down on basic training time helps with the 72-month cycle in terms of opening up more weeks to complete other tasks (maintenance, integrated/advanced phase training, surge deployments) and time for OPTEMPO-required sustainment.

Training for Hull Swaps

The Navy has already established guidelines for training needed in the event of swapping a crew from one hull to another. The first guideline is that when crews transfer from one ship to another, mission area certifications transfer with the crew. Thus, crew members will be fully certified on their current ship and will not need to repeat training they have already completed. Additionally, post–hull swap training will highlight the differences between the two ships and provide specific mission area verification. Mission area verification will be focused in the following areas:

1. Navigation (MOB-N)
2. Seamanship (MOB-S)
3. Engineering (MOB-E)
4. Damage Control (MOB-D)
5. Anti-terrorism (AT)
6. Search and Rescue (SAR)
7. Medical (FSO-M)
8. Explosive Safety (EXPSAF)
9. Amphibious Warfare (AMW)
10. Aviation (AIR) (11)
11. Communications (CCC).

Summary

Annual manpower costs for manning DDG-51 class ships is significant. The potential exists to operate the fleet with fewer crews by removing them from ships during extended depot maintenance. However, a maintenance crew is needed to maintain the safety and security of the ship and to oversee maintenance operations. The number of personnel required to stay with the ship will offset savings achieved by removing crew members.

The Navy spends significant time and resources in training ship's crews for deployments. Training time entitlements and guidelines govern the training and were used when constructing the 72-month operational cycle. The Navy has used the practice of removing crews from ships, and the general rule is that the qualifications of the individuals, teams, and the crew go with them when they transfer to another ship. Therefore, the possibility of moving crews from one ship to another is feasible, and we used it in our analysis.

CHAPTER FIVE

Effect of a Longer Operational Cycle on Personnel Tempo and Operating Tempo

A comparison of the time allocated for maintenance, training, sustainment, deployment, and other activities between OFRP and the 72-month cycle clearly shows that the 72-month operational cycle allows for more operational availability. How much more is a function of whether three or four deployments are carried out in a 72-month period. As the vessels that provide mission support to the CSG, surface combatants have very little flexibility in the duration of their deployments with the CSG, as is reflected in the new cycle. Where flexibility does come into the surface combatant schedule is in the decision of how many independent deployments to conduct and for how long.

In the 72-month operational cycle, the second and fourth deployments are independent deployments. The length of these deployments is determined by thresholds called PERSTEMPO and OPTEMPO. These thresholds exist to protect the quality of life of naval personnel: to ensure that they see their families and sleep in the own beds whenever possible.

We repeat the current OFRP schedule, which forms the basis of past PERSTEMPO and OPTEMPO calculations, in Figure 5.1, and we compare the 72-month cycle with the OFRP cycle. Note that, under the notional OFRP cycle, few deployed days and time are indicated in the final sustainment period. Again, we posit that ships will enter into an extended maintenance period prior to entering an extended (72-month) operational cycle. Additionally, we track PERSTEMPO limits for each individual, and roughly one-third of a ship's crew will rotate

Figure 5.1
36-Month Current Employment Cycle—the Optimized Fleet Response Plan

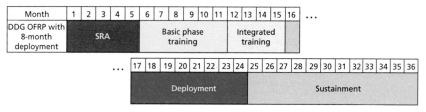

RAND *RR1235-5.1*

yearly. Therefore, the number of qualifying underway days that count against an individual's PERSTEMPO is not consistent across an entire crew (e.g., the crew that finishes the deployment at month 24 will be different from the crew that begins under a new 72-month operational cycle).

Personnel Tempo

PERSTEMPO thresholds dictate the amount of time Navy personnel can engage in official duties that make it infeasible to spend off-duty time at home. Examples of such duties would be spending time away on deployment or in training that prevents an officer or enlisted Navy member from sleeping in his or her own bed. The following PERSTEMPO thresholds apply:

1. Personnel can be deployed for a maximum of 220 days within a 365-day period. Compensation in the amount of $16.50 per day is awarded for any additional days spent away from home, whether on deployment or in training, up to $495 in a month.[1]

---

[1] In late September 2014, the Navy began paying sailors Hardship Duty Pay–Tempo (HDP-T). The basic requirement is that HDP-T is authorized for crew members that are on deployment lasting more than 220 consecutive days. Sailors are paid $16.50 per day over the 220-consecutive-day threshold, with a maximum $495 payment per month. The Office of the Secretary of Defense and the Secretary of the Navy approved the policy on September 17, 2014, and Navy guidance on this policy is contained in MILPERSMAN 7220-075, 2014.

2. Personnel can be kept away from home for a maximum of 400 days in a 730-day (two-year) period.

Exceeding either of these two thresholds requires advanced approval from a first flag officer in the chain of command or the Secretary of Defense in the event the 2001 National Security Waiver is lifted.

It should be noted that the Navy's current 36-month schedule (as well as the previous 32-month schedule) violates PERSTEMPO guidelines. In both cycles, ships undergo eight-month deployments (240 or more days deployed), which exceed the 220 days in a 365-day period just for deployment. Preparation for deployments includes training that also keeps personnel from home, on average 24 days per quarter, which increases the number of days engaged in official duties that prevent one from sleeping in his or her bed above the 220-day limit.

Operating Tempo

Navy leadership has determined that, to meet GFM presence requirements and maintain stability in training and maintenance, it must plan for CSGs and certain other units to conduct deployments of approximately eight months in length (OPNAVINST 3000.13D, 2014). OPTEMPO and PERSTEMPO limits are driven by legislation (10 U.S.C. 991). Guidelines for OPTEMPO—the rate at which units are involved in military activities away from homeport or from a permanent duty station (in other words, time spent on deployment)— also restrict how long the second and fourth deployments can be in the 72-month operational cycle. OPTEMPO guidelines are as follows (OPNAVINST 3000.13D, 2014):

1. Deployments can be a maximum of 245 days.
2. The optimum ratio of time spent deployed to time at home— known as the turnaround ration—is 1:2, with a minimum of 1:1, for all active component units. CNO approval is required for any ratio below the minimum.

3. The maximum cumulative number of days personnel can spend deployed is 540 in a 1,095-day (three-year) period. The 1,095-day period is determined by looking at deployment history of the previous two years and the planned deployment for the upcoming year.

4. Should a potential GFM solution require exceeding the OPTEMPO control levels listed above, CNO approval is required before presenting the proposal to the Joint Staff.

Start Times of Second and Fourth Independent Deployments in 72-Month Cycle

The previous discussion of PERSTEMPO thresholds demonstrates that designing a new cycle that increases operational availability while respecting PERSTEMPO thresholds is difficult, because the lengths of the first and third deployments, which align with the CSG, limit the duration of the second and fourth deployments. In addition, PERSTEMPO thresholds dictate when the second and fourth deployments can begin. As shown in Figure 5.2, meeting the dwell goal after a seven-month CSG deployment prevents the independent deployments from beginning sooner than seven months after the deployments with the CSG end. To meet the Navy's 1:2 dwell ratio goal, the second and fourth deployments should not begin less than 14 months after the first and third deployments end.

To meet the dwell goal with seven-month initial deployment, the second deployment must start 14 months after end of the first deployment and would be limited to approximately 2.5 months to remain compatible with CSG deployments.

A second deployment in 36 months could start sooner/last longer by deploying before the 2.0 dwell goal (and above the minimum) and still maintain the ship's alignment with next CSG deployment.

As can be seen from this discussion, the length of the second and fourth deployments is closely linked to the deployments conducted with the CSG. PERSTEMPO and OPTEMPO restrictions limit use of the surface fleet for independent deployments. On the other hand,

Figure 5.2
Turnaround Ratio, by Deployment Length

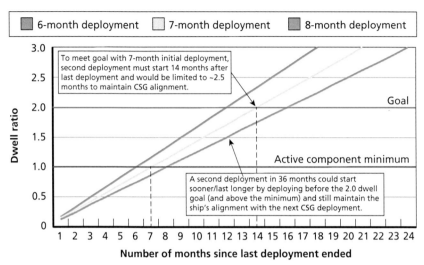

RAND RR1235-5.2

these independent deployments build flexibility into the operational availability of ships, because decisions about the second and fourth deployments can be made based on personnel needs, operational demands, and other issues. The overall effect on retention is unknown but is something the Navy should pay close attention to and adapt to accordingly.

Deployment Durations in the 72-Month Operational Cycle

We attempted to maintain PERSTEMPO and OPTEMPO thresholds when developing the 72-month operational cycle, though doing so proved difficult. Seven-month deployments with the CSG inevitably violate PERSTEMPO thresholds of 220 days spent away from home in a 365-day period, though less egregiously so than OFRP, which has eight-month deployments. Figure 5.3 illustrates how many cumulative underway days personnel spend away from home in a rolling 12-month period. The blue dotted line indicates the PERSTEMPO threshold,

Figure 5.3
RAND Calculations of Count of PERSTEMPO Days for Rolling 12-Month Period—OFRP and Potential 72-Month Operational Cycle

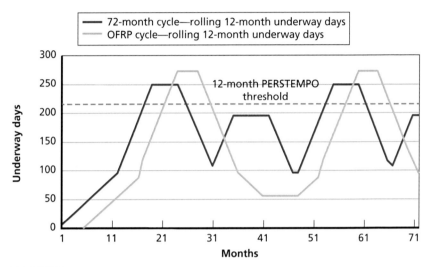

RAND *RR1235-5.3*

the dark purple line represents the 72-month operational cycle, and the light purple line represents the OFRP cycle. As can be seen by the difference between the peaks of the underway days for the 72-month operational cycle and OFRP, the new cycle violates PERSTEMPO thresholds less than OFRP.

The 72-month operational cycle also violates PERSTEMPO thresholds for a 730-day period, while OFRP does not. Figure 5.4 illustrates how many cumulative underway days sailors spend in a 24-month period. The red dotted line represents the PERSTEMPO threshold, the dark purple line represents the 72-month operational cycle, and the light purple line represents OFRP.

Figure 5.5 shows the relationship between seven-month deployments with the CSG and 4.5-month independent deployments and PERSTEMPO thresholds. The 12-month threshold and 24-month thresholds are exceeded, but the rolling three-year threshold is not. This cycle was constructed to maintain alignment with the CSG. We illustrate two seven-month CSG deployments and follow-on 4.5-

Figure 5.4
RAND Calculations of PERSTEMPO Days for Rolling 24-Month Period—OFRP and Potential 72-Month Operational Cycle

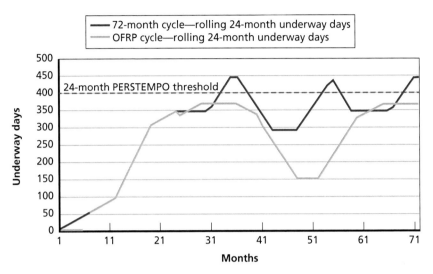

RAND *RR1235-5.4*

month independent deployment in Figure 5.5 to evaluate the impact on OPTEMPO and PERSTEMPO thresholds.

Figure 5.5 indicates that second and fourth deployments longer than 4.5 months would violate all PERSTEMPO thresholds. Thus, there are no circumstances under which the independent deployments can be longer than 4.5 months without violating PERSTEMPO. In fact, we found that the independent deployments could only be 2.5 months in duration to ensure that no PERSTEMPO thresholds are exceeded. This limits the Navy's flexibility in the duration of the independent deployments. Depending on the deployment location, when transit time is considered, a ship's on-station time for the second and fourth deployment in this cycle would be considerably limited.

Figure 5.5
RAND OPTEMPO and PERSTEMPO Calculations of Notional 72-Month Operational Cycle, with Crews Hull-Swapping During Maintenance Period

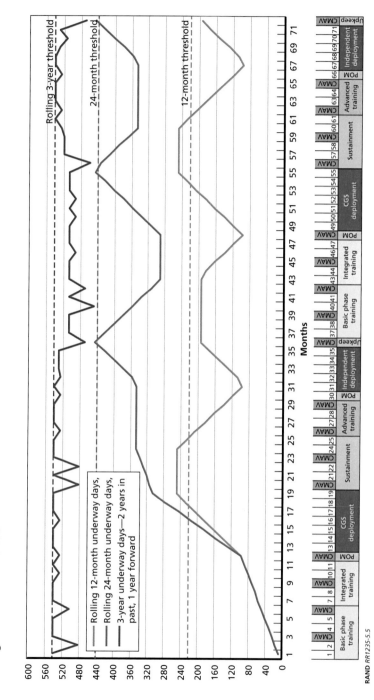

Summary

OPTEMPO and PERSTEMPO restrictions limit use of ships during an operational cycle. While a ship may be materially ready to deploy and the crew may be qualified, tempo restrictions limit the ship's underway time.

The analysis indicates that the first and third deployments with the CSG and the length of these deployments (anticipated to be shorter—seven months—in the future) affect and dictate the start time for the second and fourth deployments in the 72-month cycle. The constraint that DDG-51s must remain aligned with the CSG for the third deployment limits the second deployment's length. The requirement for depot maintenance at the end of the 72-month cycle limits the length of the fourth deployment.

Our analysis indicates that enough flexibility exists with the second and fourth deployments' duration (i.e., shorter deployments) to achieve tempo limits. The 72-month operational period is highly coupled, and changes or delays early in the schedule can have a "domino effect" on ship employment events later in the cycle.

Deployments can increase, but the result is increased underway time, away from home, that requires careful management to stay within PERSTEMPO and OPTEMPO guidelines. The DDG-51 alignment to a CSG and corresponding deployments drive underway days and impose constraints on additional deployed operations in a 72-month cycle. Finally, increased underway time will likely negatively affect retention. The potential effect of increased PERSTEMPO/OPTEMPO in a 72-month cycle on personnel retention is to be determined.

CHAPTER SIX

Scheduling Model and Its Outputs

To explore fully the implications of extending the deployment cycle of the DDG-51 surface fleet, and to gain a better understanding of the dynamics involved in swapping crews between ships, we developed a fleet maintenance-scheduling model. The motivations for developing such a model were to attempt to define a number of the limitations, challenges, and constraints discussed previously in this report; quantify how these would affect the schedule of each ship; and consequently determine the amount of deployment and operational availability the fleet could achieve as a whole. Consistent with the goal of getting the maximum amount of deployed time out of each ship, we developed a model that would maximize deployed time, subject to our constraints.

This chapter describes the model and the assumptions that underpin it. In it, we also present the model results, which make it possible to track all the ships in the class as they move out of the shorter cycles and into the 72-month cycle. We also track the time that a ship either is deployed or can be deployed. Finally, we present the cost implications of moving to the 72-month cycle, specifically noting whether any cost savings might accrue. Information about the model design appears in Appendix B.

Assumptions

To proceed with our model, we had to make a number of assumptions. Changing any of these assumptions has the potential to alter the outcome of the modeling. Nonetheless, all of our assumptions were informed by policy, guidance, subject-matter expert interviews, and the judgment of the study team, and they were vetted by the sponsor.

Many of these were discussed in Chapter Three in the description of the new maintenance demands, but there are others as well. The assumptions most germane to the modeling are listed in Table 6.1.

We use the 32-month cycle as the cycle from which ships transition, to remain consistent with the basis for new maintenance demands. The fleet is currently in the process of transitioning to the 36-month OFRP cycle. However, the DDG-51 TFP describing maintenance requirements had not yet been developed and published at the time of our analysis, and thus we used the 32-month cycle TFP as our basis.

Design of the Model

We wrote the fleet maintenance scheduling model in the General Algebraic Modeling Systems integer programming language, and we used the Cplex solver to find the optimal solution to our objective. Within

Table 6.1
Modeling Assumptions and Potential Impacts on Results

Modeling Assumption	Impact on Results If Incorrect
Ships enter the new cycle after completing a docking availability in the 32-month cycle	Changes maintenance requirements
Availabilities are fully funded and start and complete on time	Changes all scheduling, as well as potential future maintenance requirements
Sufficient dry-dock capacity is available on both coasts	Delays possible, which affects scheduling
FDNF ships are excluded from the model	None; FDNF ships currently operate on different cycle than CONUS ships
Crews do not change homeports	Costs increase
Ships can change homeport after an availability to join an available crew, assuming that they are on the same coast (e.g., an Atlantic-based ship must remain in an Atlantic homeport)	Costs increase

our program, we can represent the ships to be in any number of states, such as maintenance, de- and recrewing, or sustainment, hence the "integer" in integer programming. Each of these states is represented by a discrete number. The objective function that our model optimizes describes the total number of cycles, and hence the total number of deployments, which are completed across the fleet through the end of each ship's ESL. By maximizing the total number of cycles completed subject to the constraints we impose, we ensure the largest number of deployments that each ship can individually complete through ESL, and thus the most time operational and deployed.

The basic construct of the model is to define the DDG-51 fleet, as is it exists today, including such pertinent information as the flight of each ship, age, maintenance requirements, homeport, and transition date into the new extended cycle. We define the new 72-month operational cycle and maintenance periods and provide them as inputs as well. We then iterate through time by month and advance each ship in its cycle, subject to the constraints we define. These constraints are (1) the maintenance requirements for each ship—when a ship reaches the end of its cycle, it must enter maintenance for a specific amount of time determined by where the ship is in its life cycle—and (2) crew availability—as the ship is decrewed at the beginning of a maintenance period, it must be recrewed upon reentry into the operational cycle, and thus a crew must be available. Should a ship encounter a constraint, it is placed into sustainment until that constraint has been lifted. This is the parameter that allows the otherwise rigid cycle to vary.

When a ship enters maintenance availability, its crew joins a queue awaiting the next ship exiting its maintenance period, and the crew reenters the employment cycle once a ship comes out of maintenance. Once a ship that needs a crew becomes available, the crew at the top of the queue reports to that ship. We program a one-month buffer for this to take place. We track crews using the ships they are originally crewing. That is, DDG-51 is crewed by Crew-51. However, at later points in time, ships and crews are not matched as they were initially. That is, DDG-51 may be crewed by Crew-74, or some other crew as dictated by the model, in the future. The model allows us to "turn off" specific crews at specific points in time, such that we can test whether the removal of a crew will affect operational availability.

Once a ship hits its retirement date, it is removed from the fleet, and the ship and its accompanying crew is effectively "turned off" in the model. The model runs through the retirement of the most recently commissioned ship in the fleet, DDG-112. DDG-112 was commissioned in October 2012 and is a Flight IIA ship with an ESL of 40 years. Thus, our model runs until October 2052. The model produces a month-by-month breakdown of each ship, including its life-cycle phase and which crew is currently onboard. Laying out the entire fleet monthly, we are able to determine how many ships are in any of the given states simultaneously, including how many are deployed, how many are in maintenance, and how many man-days of maintenance are to be executed in that month.

Extending the time between major maintenance availabilities requires an increase in maintenance to be done between these availabilities. We posit that the increase in man-days will be similar to the number of man-days performed on DDG-51 ships in the Navy's FDNF, as per Table 6.2. In addition to increasing the number of maintenance man-days for CMAVs, the new cycle will succeed only if the type of work conducted during CMAVs focuses on life-cycle critical maintenance, such as tank and uptake work, rather than cosmetic repairs. This will ensure that maintenance critical for ensuring that ships reach their ESL is conducted both inside and outside regular depot availabilities, when most of that work is conducted in OFRP. Conducting life-cycle maintenance during CMAVs is crucial in the 72-month operational cycle, because regular depot availabilities will occur only every six years. Maintenance issues will inevitably arise that, if not addressed

Table 6.2
Current and Proposed Annual Continuous Maintenance Availability Man-Days for DDG-51-Class Ships

	Flight I/II	Flight IIA	Flight IIA (FDNF)
Current	1,700	1,900	3,500
Proposed	3,400	3,800	—

quickly, will detract from a ship's ESL. The pace of addressing life-cycle critical maintenance must be constant, and maintenance should not be deferred until the next depot availability. The 72-month operational cycle model allocates the maintenance man-days for this to be conducted during CMAVs.

One issue that arises with increasing the number of maintenance man-days for CMAVs is whether they should be longer to give maintenance providers time to finish the work or whether the rate of work executed during a CMAV should double, meaning the number of people working on a CMAV increases. Doubling the rate of work is somewhat complicated, because space aboard a vessel is limited, and workers compete for the same space to conduct their tasks. The Navy should address the issue of how larger CMAVs should be conducted with maintenance providers.

Model Results

The output of the model is a month-by-month breakdown of each ship's state in every month. A small example of what the output looks like can be seen in Figure 6.1.

This view permitted us to sum the entire fleet each month to determine the number of ships in any of the given states for four cycle options: 32-month, 36-month, and two variants of the 72-month option, one with three deployments and one with four. At the request of the sponsor, we considered the three-deployment option as a possible solution to the OPTEMPO and PERSTEMPO constraints. In addition, we tracked which type of availability each ship was in in a given month to determine the number of man-days executed that month and, by extension, an estimate of how much that maintenance would cost.

Figure 6.2 describes how much time the entire fleet spends in each state throughout the rest of the fleet's life. In addition, we track operational availability (A_o), which is the sum of sustainment, POM, and deployment periods. The sum of these three periods is the time that a ship either is deployed or can be deployed. This sum appears in the

Figure 6.1
Sample Model Output

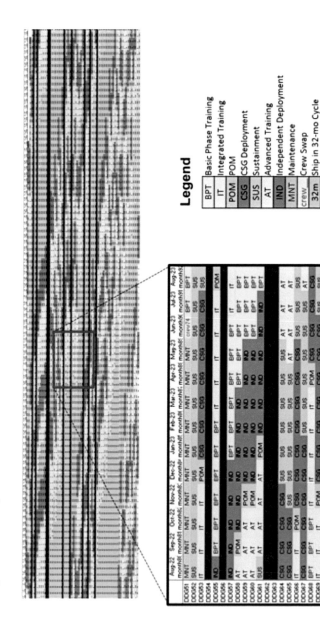

RAND RR1235-6.1

legend at the top of Figure 6.2 as a percentage of total time. Because the time scale for the employment options are relatively long—anywhere from slightly less than three years to upward of seven—we calculate these times across the ship's entire life. Limiting the timespan over which we make this calculation has the potential to skew our results. For example, if we consider only the next ten years, the 72-month operational cycles would appear to have the fleet in maintenance for an even shorter amount of time, because the majority of the fleet would not have completed full maintenance availability within the ten-year window.

Operational availability does not vary greatly across all four employment options. It is lowest at 50.0 percent for the 32-month cycle found in the TFP we used to determine maintenance demands, and highest at 56.8 percent in the 72-month operational cycle with three deployments. The 72-month operational cycle with four deployments has a lower A_o because the ship must spend more time in training to correspond with more deployments. Between the three- and four-deployment 72-month operational cycles, deployed time is 5.7 percent

Figure 6.2
Fleet-Wide Time Spent in States for Differing Employment Cycles

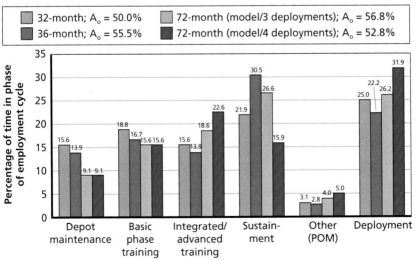

■ 32-month; A_o = 50.0%   ■ 72-month (model/3 deployments); A_o = 56.8%
■ 36-month; A_o = 55.5%   ■ 72-month (model/4 deployments); A_o = 52.8%

greater, at 31.9 percent in the four-deployment case, because of the additional deployment. This gain comes at the expense of sustainment time and also requires more time spent in integrated and advanced training, as previously mentioned.

The number of ships simultaneously in maintenance is another important consideration when comparing the costs and benefits of transitioning to a longer operational cycle, because it relates directly to the number of crews that would be available to move to another ship. There are fewer availabilities for a ship in the 72-month operational cycle, though their duration is longer than even a typical docking availability in the 32-month cycle. In Figure 6.3, we track the number of ships simultaneously in maintenance by coast. Figure 6.3 is intended as a companion display to Figure 3.1, which shows the number of ships in maintenance and as a subset ships in docking availabilities, under the 32-month cycle. We plot a 12-month moving average to aid in visualizing fleet-wide trends, which are not obvious when viewing the un-averaged data.

Figure 6.3
Number of Ships Simultaneously in Maintenance, by Coast (12-month moving average) in the 72-Month Cycle (modeled)

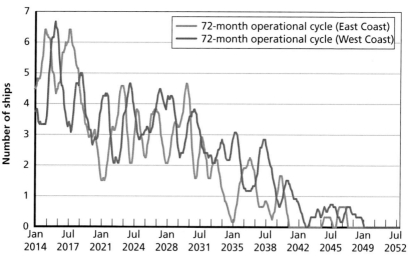

The number of ships simultaneously in maintenance levels off in the early 2020s through the end of the decade. This conforms with our expectations, because the final ship in the DDG-51 fleet transitions to the new operational cycle in FY 2022 in our model run, marking the end of availabilities under the 32-month cycle. Finally, DDG-51 is the first ship in the fleet to retire and is set to do so in FY 2026. Our model corroborates this result because, as Figure 6.3 shows, the number of ships in maintenance begins to tail off at the end of the 2020s, which coincides with a reduction in fleet size as ships reach the end of their ESL.

Cost Implications

The primary motivation for this study was to explore the possibility of getting more use out of the DDG-51 fleet and to illuminate issues and challenges that the Navy must grapple with should it decide to proceed with this new operational cycle. A secondary motivation is to save costs. These potential cost savings would result in a reduction in the number of DDG-51 crews. As ships enter maintenance, they are decrewed, and as long as there are ships in maintenance under the extended cycle, there will be surplus crews. If the Navy were to divest of these surplus crews and were to reduce end strength to reflect this divestment, there would be a potential for cost savings. It is important to note that any reduction in crews will manifest in cost savings to the Navy and U.S. government only if end strength is reduced. Should the crews be divested but placed elsewhere in the Navy, no cost savings can be realized.

There are three components to determining cost savings resulting from crew reductions. First, there is the number of crews that can be removed from the force structure. Because the primary motivation of this study is to use our capital assets to the greatest extent possible, we allowed crews to be cut from the structure only if doing so would not affect operational availability in the future. That is, as long as crews are removed from ships, there will be a mismatch between the number of crews required to operate the fleet and the number of ships. There will be times when there is either a crew without a ship or a ship without a crew. In keeping with the goal of this report to get maximum use from the Navy's capital assets, we assumed that the Navy never loses a ship because no crew is available. This means that there will be times when crews are available to staff a ship but do not have an available ship to staff. It is important to note that relaxing the restriction never to lose ship operational availability would alter the output of our analysis. There would likely be more crews that could be reduced if this were the case.

Second, we remove only a fraction of the crew because shore safety and oversight must be maintained while the ship is in its availability. This issue is discussed in detail in Chapter Four. Third, the annual cost of a DDG-51 crew is the last piece necessary to determine cost savings that might result from crew reductions. Crew costs are also discussed in Chapter Four. We use a range of costs for a DDG-51 crew—the smaller cost is the cost to the Navy, while the larger cost is that to the U.S. government and encompasses benefits and entitlements not covered by the Navy.

Table 6.1 displays our results. In the table, the numbers highlighted in green in the annual savings row give ranges: The leftmost number is the savings to the Navy, and the rightmost number is the savings to the federal government.

We determined that a total of two DDG-51 crews could be reduced in FY 2022. This result is consistent with the results of our modeling. The first ships enter the longer operational cycle in FY 2016, and thus enter their first availability six years later in FY 2022. Because these are the first ships to decrew during maintenance, their crews do not have another ship waiting for them to staff. Therefore, the first ships to enter maintenance would be the ones whose crews would be divested. As previously mentioned, this number would likely increase should we relax the assumption to not lose operational availability because of a lack of crews. The two divested crews would translate to an annual savings of $45 million to the Navy and $66 million to the government, provided that there would be a commensurate reduction in Navy end strength. This is due to the fact that as two crews are to be removed, approximately 50 personnel or one-sixth of each must remain with the

Table 6.1
Change in Number of Crews Required to Operate Fleet Under a 72-Month Operational Cycle, by Fiscal Year
# of Reduced Crews × Fraction of Crew Removed × Annual Cost of Crew = Potential Annual Savings

	FY 2016	FY 2017	FY 2018	FY 2019	FY 2020	FY 2021	FY 2022	FY 2023	FY 2024	FY 2025	FY 2026
Cumulative # crews reduced	0	0	0	0	0	0	2	2	2	2	2
Max annual total crew savings ($ millions)	0	0	0	0	0	0	<45–66	<45–66	<45–66	<45–66	<45–66

NOTES: Costs are in constant FY 2014 dollars. A range of costs are shown for crew savings and total (crew and availability) cost changes. This range represents savings to the Navy on the low end ($45 million per year) and savings to the government as a whole on the high end ($66 million per year). The difference accounts for benefits and entitlements that the government provides.

ship as caretakers. Therefore, five-sixths of each crew will be removed, equating to crew costs of $26.9 million and $39.7 million annually to the Navy and government, respectively, as described in Chapter Four. These values used with the equation in Table 6.1 provide this range.

Savings from crew reductions is not the only component of cost that we considered. Maintenance demands also change under the new operational cycle, and we used our model output to capture the effects on cost. There are two aspects to consider: (1) the cost of additional CMAV man-days under the extended cycle and (2) the change in cost of CNO availabilities between the fleet remaining on the 32-month cycle and the fleet transitioning to a 72-month operational cycle. These aspects are displayed in Table 6.2. The figures in green are savings; those in red are cost increases.

Additional CMAV man-days were calculated by determining when a ship transitioned to the new cycle and what flight the ship is, then taking the total amount of additional man-days the ship is to receive across a year and dividing it evenly among every month in the year. We then applied this average monthly amount to ships in the new cycle when they were not in maintenance availability. The distinction between flights of the ship is necessary because Flights I/II and IIA receive differing amounts of additional man-days.

We calculated the change in CNO availability costs by determining the total man-days to be executed in CNO availabilities across the entire fleet when the fleet stays in a 32-month cycle and when the fleet transitions to the 72-month operational cycle. We used the age of each ship to determine when it would be in specific availabilities in both the old and new cycles. We assumed that a man-day of effort in a private yard is equivalent to $500, and we used this value as our conversion factor from man-days to dollars.

Again, these results are consistent with the design and assumptions of our model. Ships begin to transition to the new cycle in FY 2016, and thus we begin to see the cost of additional CMAV man-days appear at this point and grow year after year until FY 2022, when the last ship in the fleet transitions to the extended cycle. From FY 2022 on, the exact amount of additional CMAV man-days varies as the number of ships in availabilities, and thus not receiving continu-

Table 6.2
Changes in Maintenance Cost Under a 72-Month Operational Cycle, by Fiscal Year

	FY 2016	FY 2017	FY 2018	FY 2019	FY 2020	FY 2021	FY 2022	FY 2023	FY 2024	FY 2025	FY 2026
Cost of additional CMAVs in new cycle ($ millions)	>2.1	>5.5	>11.1	>16.7	>24.2	>30.6	>36.8	>40.0	>41.4	>43.1	>40.1
Change in CNO availability cost ($ millions)	0	0	<11.4	<32.1	<57.0	<86.7	<18.7	>130.6	>106.8	>67.5	>164.6
Total change in maintenance costs ($ millions)	>2.1	>5.5	<0.3	<15.4	<32.8	<56.1	>18.1	>170.6	>148.2	>110.6	>204.7

NOTE: Costs are in constant FY 2014 dollars.

ous maintenance, varies. The change in CNO availability costs shows savings from FY 2018 through FY 2022 as ships in the new 72-month operational cycle bypass the availabilities they would enter were they still in the 32-month cycle. However, in FY 2022, ships in the fleet begin to enter their first availability in the new cycle, and these savings turn into additional costs by FY 2023. The new maintenance package sizes are very large in size and duration, so the additional costs relative to the 32-month cycle are substantial.

The cumulative effect of these two maintenance components leads to a slight increase in costs during the first couple of years after ships begin to transition to the extended cycle, followed by four years of cost savings because of the lack of maintenance availabilities the fleet would undergo in the 32-month cycle, and finally an overall increase in cost because of the additional CMAV man-days and increase and size and duration of maintenance availabilities in the 72-month cycle.

Table 6.3 displays the combined maintenance and crew reduction, using the same color convention as in Table 6.2. It shows that when the net effect of fewer crews and increased maintenance cost is taken into account, the shift to the 72-month cycle costs the Navy more than the shorter deployment cycles. Thus, from a cost perspective, there is no reason to go to a longer cycle.

From FY 2023 through FY 2026, the increase in costs will be anywhere from $44.6 million to $159.7 million in a given fiscal year. However, this analysis was conducted to satisfy our primary motivation for a longer employment cycle and does not affect the results from changes to operational availability described earlier in this chapter, nor does it affect any of the issues, challenges, and additional risks that the Navy would inherit should it choose to extend the operational cycle of the DDG-51 fleet. That is, the Navy must decide whether the increased operational availability of the fleet is worth the increase in costs the Navy would incur by going to a 72-month cycle.

Table 6.3
Combined Maintenance Cost and Crew Reduction Savings Under a 72-Month Operational Cycle, by Fiscal Year

	FY 2016	FY 2017	FY 2018	FY 2019	FY 2020	FY 2021	FY 2022	FY 2023	FY 2024	FY 2025	FY 2026
Total change in maintenance costs ($ millions)	>2.1	>5.5	<0.3	<15.4	<32.8	<56.1	>18.1	>170.6	>148.2	>110.6	>204.7
Max annual total crew savings ($ millions)	0	0	0	0	0	0	<45–66	<45–66	<45–66	<45–66	<45–66

NOTES: Costs are in constant FY 2014 dollars. A range of costs are shown for crew savings and total (crew and availability) cost changes. This range represents potential annual crew savings to the Navy on the low end ($45 million per year) and savings to the government as a whole on the high end ($66 million per year). The difference accounts for benefits and entitlements that the government provides. The total change in costs reflects the combination of change in maintenance and crew savings.

CHAPTER SEVEN

Findings, Conclusions, and Recommendations

This chapter sums up the findings from our research and what we conclude based on those findings.

Findings

Operational Cycles

Overall, we find that the Navy has experimented with operational cycles before and that the number of months in the cycle has changed several times. The fleet is currently transitioning to the OFRP 36-month cycle, and it will take time to move all ships to this new cycle. The OFRP cycle is planned to have a single eight-month deployment over the 36 months, with deployed operations making up 22 percent of the cycle length. Compared with different employment used in the past, deployed operations have generally been greater in shorter cycles. Options do exist to increase deployments in the longer operational cycle examined in this analysis.

Maintenance

The DDG-51 TFP details maintenance requirements needed to reach ESL for these ships. Our research indicates that maintenance planning and execution is challenging to accomplish today, and data indicate that maintenance execution falls below the requirement. The Navy is developing a 36-month OFRP maintenance plan, but it is not yet approved. In this study, we used the 32-month maintenance requirements in the 32-month TFP. While SURFMEPP is beginning to document maintenance accomplishment, it is at the outset of its work. It will take time to document and analyze ship maintenance requirements and execution.

Navy senior engineers have concerns with an extended interval between maintenance. The risks in moving to an extended interval include the unknown effect on maintenance and modernization requirements, private maintenance industrial base, current challenges, and future maintenance funding.

Training and Manpower

The amount of time needed to perform training in a ship's employment schedule is considerable. We find that ships must undergo ULT every 36 months. Personnel turnover is a major driver of the need for ULT. A crew's qualifications remain with them during a hull swap. These crew qualifications and cohesion would potentially reduce training demands compared with the complete outfitting of a ship with a new crew. While hull swapping is an alternative, configuration variances exist between ships of a given class, and additional training would be needed in the new hull. Ship crews must receive advanced training for the second and fourth deployments in the 72-month cycle. Additional certification training is needed based on the missions the ship would perform in these second and fourth deployments.

The annual cost of manpower for a DDG-51 is significant and ranges from $29 million to $43 million. The potential does exist for removing crew by means of hull swap and operating class with fewer crews. However, increased costs will result from shifting work normally done by the crew during an availability to the private maintenance provider. Potential crew savings from decrewing during maintenance depend on the Navy's planned use of the crew that was removed. If the Navy assigns the departing crew to other Navy billets with no resulting decrease in end strength, then no cost savings will be realized. If, however, the Navy plans to operate the fleet with fewer crews and plans accordingly to reduce end strength, then manpower savings can be achieved.

Operating Tempo

The Navy exceeds PERSTEMPO limits today with eight-month deployments in the 36-month cycle. When predeployment nondeployed underway time is factored and calculated, ships in the OFRP cycle exceed the 12-month PERSTEMPO thresholds of 220 days. Eight-month deployments equate to more than 240 days. The Navy currently compensates sailors with "hardship duty pay" for high-tempo "operational deployment length" in excess of the 220-day threshold. However, this tempo pay does not take into account the predeployment underway time that sailors experience prior to deployment. Factoring in eight underway days per month prior to a deployment (nondeployed underway days for four months = 32 days) and an eight-month deployment (240 days), sailors could be away for home 272 (240 + 32) days in a 12-month period. The nondeployed underway time is not addressed by hardship duty pay.

As is the case today, the underway time in the 72-month cycle constructed in this analysis also exceeds the 12- and 24-month PERSTEMPO thresholds. The analysis sought options to maximize deployed time within the 72-month operational cycle. *CSG alignment and tempo* policy limits deployment time in an extended cycle.

Conclusions

A 72-month operational period increases operational availability of DDG-51s from 50.0 percent to 56.8 percent, corresponding to a 13.6 percent increase in the number of operational ships. This cycle affects critical factors for effective ship operations, including maintenance, manpower, training, and tempo of operations. Overall, we found that maintenance challenges exist now, that an extended cycle injects greater uncertainty into current maintenance challenges, and that changes are needed to correct inefficient maintenance approaches that are used today.

The normal rotation of personnel dictates need for ULT twice in the longer 72-month cycle. Removal of the crew through hull swapping has advantages, because it maintains crew cohesion, continuity, and qualifications.

The training under the 72-month operational period would follow the normal CSG sequence for the first and third deployments in the cycle, with the second and fourth deployments requiring tailored integrated and mission area training recertification.

OPTEMPO under this extended cycle tightly constrains scheduling. Slips in scheduling will greatly affect follow-on training and maintenance events. Moreover, this compressed schedule and underway time presents a limited duration for the second and fourth deployments. Current deployments surpass PERSTEMPO thresholds, and so will the second and fourth deployments in the 72-month cycle.

Costs will increase from delaying maintenance, from transferring ship's maintenance to a private depot provider, and from additional crew that are needed ashore to support the ship. Potential cost savings by removing the crew during maintenance are offset by these increased costs. The opportunity to achieve savings through crew removal is limited.

Recommendations

If the Navy is interested in increasing operational availability of ships by extending the maintenance interval for DDG-51s, it should implement the following recommendations. We have divided them into two categories: (1) maintenance planning and execution and (2) training and operations.

Maintenance Planning and Execution

1. Before going to a longer interval between depot maintenance, the Navy should correct impediments to availability execution.
2. Determine maintenance requirements. Senior Navy Engineering Duty authorities indicated that the Navy has not fully identified and documented the conditions of surface combatants,

particularly the condition of tanks. Tank maintenance is a major driver of maintenance and funding needs for depot work.

3. Develop a maintenance plan for the longer cycle. Navy maintenance authorities need to develop a plan that addresses the timing and sequence of maintenance in a longer operational cycle.

4. Increase continuous maintenance man-days; focus on life-cycle critical maintenance. With a longer interval between a dedicated depot availability, increased continuous maintenance is needed to address both emergent maintenance demands and life-cycle critical maintenance.

5. Resource maintenance demands. Review of maintenance execution compared with the maintenance requirements contained in the DDG-51 TFP indicates that a ship's depot maintenance is funded below the requirement. The Navy should determine whether the TFP requirement is actually the requirement and either fund it accordingly, adjust the requirement, or determine whether the risk (of not achieving ESL) is acceptable to fund maintenance below the requirement.

6. Improve current maintenance planning and execution. Senior Navy maintenance experts indicate that current maintenance planning and execution are not as efficient and effective as they should be.

7. Evaluate effect of maintenance demands on private providers. Little data are available that address the private supply of labor or the effect a different maintenance cycle would have on the private providers of maintenance.

Training and Operations

1. If the Navy opts for a 72-month cycle, require ships to enter the cycle after CNO docking and in a high state of material readiness—senior maintenance authorities all voiced that ships must be in the highest state of material readiness to enter a cycle that requires a longer interval between depot maintenance periods. Moreover, a docking should precede this longer interval.

DDG-51s are required to be docked at an eight-year interval. Exceeding that interval would raise the risk of catastrophic and costly failure of system components that can be maintained only in a docking availability.

2. Complete evaluations (dry-docking, tank conditions) of ship material readiness. SURFMEPP indicated that an evaluation of just tank conditions would not be completed until end of FY 2016, and the repair/maintenance of the tanks would be completed in FY 2022.

3. Award CNO availabilities in a fashion that allows for sufficient time for planning the work; the surface type commander must commit funding at the time of the award.

4. Fine-tune training demands and tailor to additional deployment needs. A new operational use of ships with an independent second and fourth deployment in a 72-month cycle will increase training certification requirements for these additional deployments. Tailoring of training to meet the mission requirements of these additional deployments is needed.

5. Closely manage OPTEMPO. The Navy is exceeding tempo thresholds today with the current single eight-month deployments in the OFRP cycle. Increased deployments in a 72-month operational cycle increased tempo, which requires close management of tempo thresholds and goals.

6. Use the model in this report to support analysis. The program developed can support fleet-wide analysis. The model and analysis that we have developed for the examination of DDG-51 employment model can also be used for cruisers and amphibious ship. Moreover, it has the capability and could provide a fleet-wide examination of maintenance and operational deployments, and how best to manage the various factors that are affected.

Individuals Interviewed for This Report

We interviewed individuals from the following organizations and commands:

- Commander, Fleet Forces Command – N43, N1
- Commander, Navy Surface Forces, Pacific
  - N1, N3, N7, N43
- Commander, Naval Surface Forces, Atlantic
  - N43 – depot scheduler
- Program Executive Officer Ships
- NAVSEA, Surface Warfare directorate (SEA21)
- Commander, Regional Maintenance Centers
- SURFMEPP
- OPNAV – N96, N122, 120, 130
- BAE, a private yard maintenance provider
- Navy Manpower Analysis Center.

We also used pertinent Navy references to support our analysis. These data and documentation included the following:

- DDG-51 TFP (NAVSEA21, 2012c)
- OPNAV PERSTEMPO and OPTEMPO guidelines
- OPNAV maintenance guidelines
- training guidelines for surface fleet ships/detachments
- FY 2015 Navy Budget data
- Visibility and Management of Operating and Support Costs
- maintenance studies.

- Navy Manpower Analysis Center Ship Manning documents for DDG-51
- Office of the Secretary of Defense Full Cost of Manpower Tool.

Background of Surface Combatant Maintenance

OPNAV N81 tasked RAND to evaluate the effect of extending the maintenance/operating cycle from the present 36 to 72 months for DDG-51 *Arleigh Burke*–class destroyers. To evaluate the proposal, it is instructive to examine past practices, the current condition of the force, and Navy maintenance actions.

In February 2010, the Fleet Review Panel issued its final report on surface force readiness (Balisle, 2010). The Commander, U.S. Fleet Forces Command, and Commander, U.S. Pacific Fleet, requested the review. The report highlighted "numerous, well intentioned changes in material readiness related organizations, policies and process over the last decade" (Balisle, 2010, p. 7). It concluded, in part:

> In the last decade there have been many changes that have impacted surface force readiness. It appears the effort to derive efficiencies has overtaken our culture of effectiveness. . . . The material readiness of the surface force is well below acceptable levels to support reliable, sustained operations at sea and preserve ships to their full service life expectancy. Moreover, the present readiness trends are down. . . . Material readiness trends develop and evidence themselves over years vice months. . . . Accordingly, the most effective material readiness program is one that is consistently followed with small, evolutionary improvements made to it vice dramatic changes. (Balisle, 2010, p. 7)

Some of the report's findings are listed below (Balisle, 2010, pp. 4–7):

- "Surface Force Readiness has degraded over the last 10 years. This has not been due to a single decision or policy change, but is the result of many independent actions."
- "Optimum Manning . . . did not consider other factors such as maintenance requirements."
- "Further exacerbating shipboard material readiness was the decreased capability and capacity of shore intermediate repair capability."
- "CNO Maintenance availabilities were shortened from 15 weeks to 9 weeks and the Material Maintenance Management (3-M) program was scaled back."
- "Continuous Maintenance Availabilities (CMAVs) are being executed well below their capacity."
- "There is a growing backlog of off-ship repair requirements; this represents a large deep maintenance requirement that has not been adequately identified or resourced."
- "Shipboard distributed systems such as chilled water systems, and fire mains, structure, tanks and voids are in wide disrepair throughout the Surface Force."
- "The Panel feels there is a need to formalize a recurring third party led assessment process to properly and fully identify and manage the deferred maintenance requirement of the Surface Force."
- "The effort to derive efficiencies has overtaken our culture of effectiveness."
- "The material readiness of the Surface Force is *well below acceptable levels to support reliable, sustained operations at sea* [emphasis added] and preserve ships to their full service life expectancy."

Since issuance of the Balisle Report, the Navy has taken several steps to correct and reverse the trends. These actions will be discussed later. However, the principal questions are

- How effective have these changes been to correct and reverse these trends?
- What is the current state of Surface Force ship material readiness?

- What is the likely effect of the proposed 72-month cycle on the ships' material readiness?

U.S. Government Accountability Office Review of Surface Combatant Maintenance

To answer the above questions, we examined a number of documents and interviewed individuals and organizations familiar with ships' material readiness condition.

U.S. General Accountability Office Report and Interviews

In September 2012, approximately two and a half years after the Balisle Report, the U.S. General Accountability Office (GAO) published a report titled *Navy Needs to Assess Risks to Its Strategy to Improve Ship Readiness* (GAO, 2012). The GAO was directed to assess Navy initiatives for improving the material readiness of the Surface Force ship.[1] The report noted that in the wake of the Balisle Report, the Navy took a number of steps to bring about a more systematic and integrated approach to improve material readiness. Some of these actions were the following (GAO, 2012, pp. 13–15):

- establishment of SURFMEPP in November 2010
- establishment of Navy RMC headquarters in December 2010
- began increasing personnel manning at Navy Shore Intermediate Maintenance Activities (SIMA) in June 2011
- established a new strategy in the *Navy Surface Force Readiness Manual* (COMNAVSURFPACINST/COMNAVSURFLANT-INST 3502.3, 2012a) in March 2012.

Together, these actions were aimed at improving management of maintenance requirements and execution. The readiness manual strategy consolidated multiple assessments throughout the previous Fleet Response Plan into seven readiness evaluations at specific points in the then-current 27-month cycle.

---

[1] This directive was set forth in U.S. House of Representatives Report 112-78 (U.S. House of Representatives, 2011), accompanying the Fiscal Year 2012 National Defense Authorization Act (H.R. 1540) by the House of Representatives Armed Services Committee.

In evaluating the Navy's progress in employing its new strategy, the GAO report stated that implementation started in March 2012 and would not be fully realized until FY 2015 (GAO, 2012, p. 15). The Navy acknowledged that there are circumstances (higher OPTEMPO, etc.) that may delay full implementation of the strategy and the 27-month cycle. In particular, ballistic missile defense ships had a higher OPTEMPO, leading to quicker turnaround time. This rapid turnaround may lead to some ships not having time for their scheduled maintenance periods. The GAO stated, "Thus, ships with a high operational tempo that do not enter the maintenance phase as planned will have life-cycle maintenance activities deferred, which could lead to increased future costs" (GAO, 2012, p. 16).

GAO interviews at the Navy RMC headquarters resulted in RMC officials noting

> They currently lack the staff needed to fully execute the ship readiness assessments called for in the new strategy. . . . According to the officials, ship readiness assessments allow them to deliberately plan the work to be done during major maintenance periods and prioritize their maintenance funds. (GAO, 2012, p. 18)

The GAO report also stated, "The Navy has not undertaken a comprehensive assessment of the impact of high operational tempos, staffing shortages, or any other risks it may face in implementing its new readiness strategy, nor has it developed alternatives to mitigate any of the risks" (GAO, 2012, p. 18). Overall, some progress has been made to address maintenance challenges with surface combatants. However, challenges remain, and surface ship maintenance demands need oversight, staffing support, adequate resources to address demands, and time to accomplish them.[2]

The fleet maintenance officers for the Fleet Forces and Pacific Fleet, reporting on the evolution of the OFRP from the previous Fleet Response Plan (Berkey and Grocki, 2014), state that the OFRP is the

---

[2] In January 2014, the Commander, Naval Surface Forces, submitted his *Vision for the 2015 Surface Fleet* (Commander, Naval Surface Force, 2014) to the CNO. This document reported on a wide range of topics that affect the ability of the Surface Force to meet its 10 U.S.C. obli-

result of "recognition that our existing readiness generation model was unsustainable for our people and equipment." The OFRP is a 36-month cycle that aims to align the operational and maintenance cycle of CSGs, including the assigned destroyer squadrons. By aligning all units of the CSG to the 36-month cycle, the OFRP aims to maximize operational availability (A_o), synchronize personnel to the cycle with an acceptable PERSTEMPO, standardize training, consolidate inspections, and implement an integrated CSG forward deployment schedule of no more than eight months. The authors stated, "the number of ships . . . not completing depot maintenance availabilities on time was nearly 50 percent." Among the several reasons, the paper identified "unforeseen growth work." That is work not defined in the original work package. The growth work can come from several sources, some of which may result from missed or inadequate assessments, or insidious decline. The authors conclude by noting the 36-month cycle OFRP offers "more predictability than the existing model," while maximizing "operational availability . . . and stability in maintenance cycles" (Berkey and Grocki, 2014).

Current Status of Surface Combatant Maintenance

To understand the current Surface Force material readiness, we interviewed senior Navy and industry officials familiar with current ship maintenance. These included officials the private shipyard and at the NAVSEA, SURFMEPP, and RMC headquarters.

---

gations to "conduct prompt and sustained combat operations at sea in support of national interests." In the opening section, Commander, Naval Surface Force, reported

> We have made good progress since the "Take a Fix" effort and the Balisle Report, but the strong measures put in place have not had enough run-time to correct our readiness shortfalls. . . . All this means the surface force is not entering the POM-15 debates in a state of wholeness; that has yet to be achieved.

Further, the report continues,

> Processes to improve material condition assessment and maintenance execution are in place including elements of the Surface Forces Readiness manual—but we need run time and the right number of adequately trained and experienced Sailors and civilians which we do not currently have. . . . Regional Maintenance Centers are under-manned, though we are beginning to restore some manpower that was cut several years ago.

Private Shipyard Officials.[3] We interviewed private industry representatives knowledgeable with maintenance issues for both amphibious and destroyer-type ships. These interviewees have firsthand experience with maintenance availabilities of both ship types and are familiar with the findings of the Balisle Report. They identified a number of issues with availability execution affecting Surface Force material readiness.

Industry representatives[4] recognized that the steps Navy has taken and believe they are key to success in improving Surface Force material readiness. They described the formation of SURFMEPP as a crucial step in improving ship class maintenance management through work package documentation development. They also recognized SURFMEPP's role in tracking work package execution and development of "ship sheets." In addition to the SURFMEPP formation, they noted improvements at the Navy RMCs.

Industry representatives reported having seen extensive deterioration of ships' tanks, shell plating, and shafting as a result of deferring critical maintenance. Their observations confirm the conditions reported in the Balisle Report. Extending the operation cycle from 27 to 36 or 72 months, in their opinion, will further extend periods between availabilities and increase deferred maintenance, leading to further deterioration.[5] They pointed out that the current five-year docking interval will likely be extended to eight and perhaps ten years.

Industry representatives see several causes for the current degraded material readiness. Chief among these are the manner in which Surface Force ship availabilities are executed. They stated that several specific issues affect execution:[6]

- Availabilities are frequently cut short because of operations.
- Typically, Surface Force availabilities are initially funded to 60 percent of the work package.

---

[3] Interview with BAE Systems Ship Repair senior officials, Norfolk, Va., September 19, 2014 (hereafter referred to as "BAE interview").

[4] BAE interview.

[5] BAE interview.

[6] BAE interview.

- Remaining funds are authorized late in the availability, resulting in
  - overtime charges to meet schedule, increasing costs
  - availability schedule slippage
  - deferring the work to a later availability, further exacerbating material readiness problems.

While recognizing the positive actions the Navy has taken to improve readiness, industry representatives report that the above execution issues continue. It appears that, as the operations cycle increases, the Navy requires more funding for fuel and other operating needs, leading to fewer funds for maintenance and resulting in necessary work being deferred and further material readiness deterioration.[7]

Private industry representatives also pointed out that nuclear-powered ship and submarine availabilities are scheduled well in advance. The public yards have advance information on their workload. However, private yards have little or no information about expected work. Contract awards are made so close to the actual start date that sufficient planning is often not accomplished. In short, while the Navy recognizes the importance of the public, nuclear-capable shipyards and the shipbuilding mobilizations base, it does not appear to have the same understanding of the private repair industrial base.[8]

Finally, industry representatives stated that only a limited number of shipyards can complete complex repair/maintenance or modernization work. If the Navy continues worldwide high-tempo operations with the current fleet size and limited financial resources, it must recognize the repair industrial base and form partnerships.[9]

Naval Sea Systems Command. We spoke with senior officials at NAVSEA and SURFMEPP.[10] In interviews, senior Navy officials fur-

---

[7] BAE interview.

[8] BAE interview.

[9] BAE interview.

[10] Interview with NAVSEA senior officials, Washington, D.C., September 24, 2014 (hereafter referred to as "NAVSEA interview"); interview with SURFMEPP senior officials, Portsmouth, Va., October 8, 2014 (hereafter referred to as "SURFMEPP interview").

ther detailed the steps the Navy has initiated to correct Surface Force material condition. They noted several important achievements:

- SURFMEPP has been established and is being manned. It will significantly contribute to the discipline necessary to manage Surface Force material readiness.[11]
- The Surface Force is now using maintenance return data, as the Submarine Force does, to budget for maintenance requirements.[12]
- While not yet complete, SURFMEPP is compiling data on the actual ship tank conditions.[13]
- The Surface Force is now starting to buy back the maintenance backlog.[14]

However, interviewees also noted that considerable work is required to restore the deteriorated material readiness. While much progress has been made, a number of practices need to be addressed. They are as follows:[15]

- Only 60 percent of maintenance funds (1B4) are obligated in the first quarter of the year. The fleet retains the remaining funds for other contingencies, i.e., flying hours, steaming days, etc. Maintenance funds compete as part of Operations and Maintenance, Navy.
- Obligating funds late leads to increased costs (overtime), slipped schedules, or deferred maintenance.
- The Navy needs to award the Surface Force work packages much earlier and not wait until A-90 (or 90 days prior to the start of the maintenance availability) to award the work package. Both the private shipyards and Navy need time to plan the work.

---

[11] NAVSEA interview.

[12] SURFMEPP interview.

[13] SURFMEPP interview.

[14] NAVSEA interview.

[15] NAVSEA interview.

- The ship repair industrial base is important for effective material readiness. The Navy has to appreciate the ship repair industrial base, do a better job or partnering with it, and level loading the base.

Headquarters, Regional Maintenance Center. The Headquarters, Regional Maintenance Center, leads the local RMCs in developing and executing standardized maintenance and modernization processes, instituting common policies, and standardizing training in an effort to implement a consistent business model across the RMCs and, ultimately, to provide cost-effective readiness to the Navy's surface ship fleets. Navy Regional Maintenance Center is an Echelon III command that reports directly to the Commander, NAVSEA, and works closely with NAVSEA's Surface Warfare directorate and the SURFMEPP command for planning and executing surface ship maintenance and modernization.

Senior RMC personnel were provided an overview of the proposed 72-month cycle.[16] They were familiar with the Balisle Report and subsequent Navy actions to mitigate the Surface Force material readiness deterioration. They noted that while Navy initiatives (SURFMEPP, RMCs, etc.) have resulted in improved availabilities, CMAV effectiveness needs to be improved as well. Frequently, CMAV work is cosmetic (deck tile, lagging, etc.). The emphasis should be on life-cycle critical maintenance, such as Chemical Holding Tank (CHT), tank, and uptake work. A good deal of this life-cycle critical maintenance could be done in a CMAV.[17]

Senior RMC leaders reported that the Navy does a poor job of executing availabilities. There is a lag in the job start time. Increasing the time between award and start availability provides for better planning on the part of the executing activity and ensuring workforce availability.[18]

---

[16] Interview with RMC senior officials, Norfolk Va., November 14, 2014 (hereafter referred to as "RMC interview").

[17] RMC interview.

[18] RMC interview.

On transitioning to a 72-month cycle, senior RMC officials stated that modernizations will be more difficult and will result in a greater number of configurations. They will also require that every maintenance period be a docking availability. Specifically, they noted that[19]

- there is a need to address the crucial areas (tanks, etc.), some of which can only be done in a docking
- structural work and corrosion control will become more critical and require increased focus.

In considering the effect of the 72-month cycle on material readiness, senior RMC officials stated that the Navy has not been able to settle on a cycle length. As a result, it is difficult determine what is working and what is not.[20]

Summary and Conclusions About Current Maintenance Status

Our discussions with senior Navy maintenance experts and review of documentation indicate the following:

1. Surface Force material readiness has degraded over the past ten years. Readiness is below levels required to support reliable sustained operations at sea (Balisle, 2010; GAO, 2012).
2. Degraded material readiness resulted from an effort to derive efficiencies and not effectiveness (Balisle, 2010).
3. The Navy has taken a number of measures (establishing SURFMEPP, increased maintenance funding, increased SIMA manning, etc.) to mitigate the deterioration.[21]
4. As of 2014, while maintenance progress has been made as a result of Navy initiatives, additional run time is required to determine state of readiness.[22]

---

[19] RMC interview.

[20] RMC interview.

[21] GAO, 2012; SURFMEPP interview; NAVSEA interview; RMC interview; BAE interview.

[22] Commander, Naval Surface Forces, 2014; SURFMEPP interview; NAVSEA interview.

5. Increased OPTEMPO affects material readiness in two ways: It frequently cuts short maintenance availabilities and increases wear on ship systems (Malone et al., 2014).

6. Current availability practices (late availability award, focus on "aesthetic" work, etc.) frequently result in deferring crucial ship system work, increasing the maintenance backlog.[23]

7. Surface Force availabilities are typically funded initially to 60 percent of the requirement. Late funds authorization results in overtime charges, schedule slippage, and potentially deferred work.[24]

8. The Navy needs to recognize the importance of the repair industrial base and partner with industry and level load the base.[25]

9. While not yet complete, SURFMEPP is compiling data on actual ship conditions and tracking deferred maintenance.[26]

10. Evaluation of Surface Force material readiness will not complete until the 2018–2019 time frame.[27]

11. Deferred maintenance "buy back" has started, but requires additional time. Some suggest that the "buy back" time frame is 2022.[28]

12. The Navy has initiated a new 36-month maintenance cycle, the OFRP. It is the third cycle since the Balisle Report. The purpose is to align the operational and maintenance cycle of the CSGs.

13. The Surface Force needs stability in the maintenance cycle (i.e., cycle length) in order to fully implement current initiatives and determine what is working.[29]

---

[23] SURFMEPP interview, NAVSEA interview, RMC interview, BAE interview.

[24] SURFMEPP interview, NAVSEA interview, RMC interview, BAE interview.

[25] SURFMEPP interview, NAVSEA interview, BAE interview.

[26] SURFMEPP interview, NAVSEA interview.

[27] SURFMEPP interview, NAVSEA interview.

[28] Commander, Naval Surface Forces, 2014; SURFMEPP interview; NAVSEA interview.

[29] As discussed during interview with RADM Berkey, CFFC N43.

We concluded that no single data source or collection of data sources definitely points to the precise Surface Force material readiness. The boundary conditions for material readiness vary from a force that is fully ready to one that is unable meet its mission. The evidence and interviews suggests that neither is the case. The evidence does suggest that considerable progress has been made in implementing Navy initiatives and that the increasing maintenance backlog has been reversed. However, it will take some time to evaluate all ships and significantly reduce the backlog.[30] Poor availability practices, reported by several sources,[31] impede the desired progress.

Most observers we interviewed recognize that stability is critical to determining what is working and what is not.[32] In the past five years, the Surface Force has experienced three different cycles. Each has sought to improve conditions through efficiencies.

The current 36-month OFRP must be given sufficient "run time" for the Navy to evaluate its initiatives and practices. In addition, SURFMEPP must complete its ship evaluation (dry-docking, ship sheets), now predicted for the 2018–2019 time frame, for the Navy to evaluate the effects of the proposed 72-month cycle. The current backlog must also be greatly reduced to bring the ships closer to the "fully ready" boundary condition. Particular attention should be placed on crucial structural areas: tanks, piping, shell plating, shafting, etc.

Most of the Navy's senior engineering duty officers we interviewed, responsible for surface ship maintenance, strongly recommend that ships entering the program be in a high state of readiness, such as a new construction ship just completing its post-shakedown availability or a ship in a high state of material readiness that has just completed a CNO docking availability.[33] We concur with that recommendation.

Finally, we conclude that many of the current practices noted by a number of sources must be improved. Improved maintenance plan-

---

[30] Commander, Naval Surface Force, 2014; SURFMEPP interview; NAVSEA interview.

[31] SURFMEPP interview, NAVSEA interview, RMC interview.

[32] SURFMEPP interview, NAVSEA interview, RMC interview.

[33] SURFMEPP interview, NAVSEA interview, RMC interview.

ning and accomplishment of deferred work can be supported by better execution practices. Improved practices include early award of availabilities, increasing the percentage of funding at time of the award, focusing CMAV work on crucial ship areas rather than cosmetic work, and prioritizing time for availabilities.

Abbreviations

A_o	operational availability
CFFC	Commander, Fleet Forces Command
CMAV	continuous maintenance availability
CNO	Chief of Naval Operations
COMPACFLT	Commander, U.S. Pacific Fleet
COMUSFLTFORCOM	Commander, U.S. Fleet Forces Command
CONUS	continental United States
CSG	carrier strike group
DMP	Depot Modernization Period
DoD	U.S. Department of Defense
DSRA	Docking Selected Restricted Availability
ESL	expected service life
FCoM	Full Cost of Manpower
FDNF	forward deployed naval force
FY	fiscal year
GAO	U.S. Government Accountability Office
GFM	Global Force Management

HDP-T	Hardship Duty Pay–Tempo
NAVSEA	Naval Sea Systems Command
NAVSEA21	Naval Sea Systems Command, Deputy Commander, Surface Warfare.
OFRP	Optimized Fleet Response Plan
OPNAV	Office of the Chief of Naval Operations
OPNAVINST	Chief of Naval Operations Instruction
OPTEMPO	operating tempo (unit)
PERSTEMPO	personnel tempo (individual)
POM	pre–overseas movement
RMC	regional maintenance center
SFWL	ship's force work list
SRA	Selected Restricted Availability
SURFMEPP	Surface Maintenance Engineering Planning Program
SURFPAC	Naval Surface Forces Pacific
TFP	technical foundation paper
ULT	unit-level training

Bibliography

Alper, Omer E., S. Craig Goodwyn, Andrew Seamans, and Robert Levy, "Ship Depot Maintenance Spending and Near-Term Readiness," Alexandria, Va.: Center for Naval Analyses, DRM-2013-U-003853-1REV, July 2013.

Balisle, Philip, *Final Report: Fleet Review Panel of Surface Force Readiness*, Washington, D.C., February 26, 2010.

Berkey, Richard D., and Alma M. Grocki, "Maximizing Fleet Readiness: The Optimized Fleet Response Plan (O-FRP) Maintenance Construct—Our Fleet Is Depending on Us," *Naval Engineers Journal*, September 2014, pp. 25–28.

Blanco, Thomas A., "Evaluation Plan for Assessing Costs of Decrewing Ships During Overhaul: Pilot Ship III—USS Conyngham, DDG 17," San Diego, Calif.: Navy Personnel Research and Development Center, June 1980.

Boning, Wm. Brent, and Hoda Parvin, "Supportable Inventories: Modeling Ship Retirements, Purchases, Maintenance, and Modernization in a Constrained Budget Environment," Alexandria, Va.: Center for Naval Analyses, DRM-2013-U-005193-1Rev, September 2013.

Boning, Wm. Brent, and Andrew M. Seamans, "The Effects of Operational Tempo on Ship Depot Requirements," Alexandria, Va.: Center for Naval Analyses, CRM D00251878.A2/Final, July 2011.

"Burke: $2 Billion Backlog in Surface Ship Maintenance Hard to Dig Out of," InsideDefense.com, March 22, 2013.

Chief of Naval Operations Instruction 3000.13D, *Navy Personnel Tempo and Operating Tempo Program*, Washington, D.C.: Department of the Navy, 2014.

Chief of Naval Operations Instruction 3000.15A, *Optimized Fleet Response Plan*, Washington, D.C.: Department of the Navy, 2014.

Chief of Naval Operations Instruction 4700.7L, *Maintenance Policy for United States Navy Ships*, Washington, D.C.: Department of the Navy, May 25, 2010.

Chief of Naval Operations Instruction 4770.5g, *General Policy for the Inactivation, Retirement and Disposition of U.S. Naval Vessels*, Washington, D.C.: Department of the Navy, April 24, 2014.

Chief of Naval Operations Note 4700, *Representative Intervals, Durations, and Repair Mandays for Depot Level Maintenance Availabilities of U.S. Navy Ships*, Washington, D.C.: Department of the Navy, 2012.

Chief of Naval Operations Note 4700, *Representative Intervals, Durations, and Repair Mandays for Depot Level Maintenance Availabilities of U.S. Navy Ships*, Washington, D.C.: Department of the Navy, 2013.

Choi, Jino, Omer Alper, James Jondrow, John Keenan, Richard Sperling, and Michael Gessner, *Improving Navy's Buying Power Through Cost Savings*, Alexandria, Va.: Center for Naval Analyses, CRM D0014799.A2/Final, October 2006.

Choi, Jino, Donald Birchler, and Christopher Duquette, "Cost Implications of Sea Swap," Alexandria, Va.: Center for Naval Analyses, November 2005.

Clark, Bryan, "Commanding the Seas: A Plan to Reinvigorate U.S. Navy Surface Warfare," Washington, D.C.: Center for Strategic and Budgetary Assessments, 2014.

Commander, Naval Surface Force, *Vision for the 2015 Surface Fleet*, Washington, D.C.: Department of the Navy, January 2014.

Commander, Naval Surface Force, U.S. Pacific Fleet, and Commander, Naval Surface Force Atlantic, *Surface Force Exercise Manual*, COMNAVSURFPACINST/COMNAVSURFLANTINST 3500.11, no date.

———, *Surface Force Readiness Manual*, COMNAVSURFPACINST/ COMNAVSURFLANTINST 3502.3, March 9, 2012a.

———, *Readiness Evaluation (READ-E) Instruction,* COMNAVSURFPACINST/ COMNAVSURFLANTINST 3500.10, 2012b.

Commander, Naval Surface Force, U.S. Pacific Fleet; Commander, Naval Surface Force Atlantic; and Commander, Navy Regional Maintenance Center, *Total Ships Readiness Assessment (TSRA) Class Matrices*, COMNAVSURFPACNOTE/ COMNAVSURFLANTNOTE/CNRMCNOTE 4700, 2012.

Commander, U.S. Fleet Forces Command, and Commander, U.S. Pacific Fleet, *Fleet Training Continuum Instruction*, COMPACFLT/ COMUSFLTFORCOMINST 3501.3D, October 1, 2012.

COMNAVSURFPACINST/COMNAVSURFLANTINST—*See* Commander, Naval Surface Force, U.S. Pacific Fleet, and Commander, Naval Surface Force Atlantic.

COMNAVSURFPACNOTE/COMNAVSURFLANTNOTE/ CNRMCNOTE—*See* Commander, Naval Surface Force, U.S. Pacific Fleet; Commander, Naval Surface Force Atlantic; and Commander, Navy Regional Maintenance Center.

COMPACFLT/COMUSFLTFORCOMINST—*See* Commander, U.S. Fleet Forces Command, and Commander, U.S. Pacific Fleet.

Congressional Budget Office, *Crew Rotation in the Navy: The Long-Term Effect on Forward Presence*, Washington, D.C., October 2007.

DePalma, Thomas, and Robert Trost, "Combat System Part Maintenance: What Are the Benefits of Technical Assistance?" Alexandria, Va.: Center for Naval Analyses, March 2012.

Department of the Navy, *FY 2015 Department of the Navy Budget, Section III: Readiness*, Washington, D.C., 2014.

GAO—*See* U.S. Government Accountability Office.

Gortney, Bill, *Optimizing the Fleet Response Plan (Brief)*, January 15, 2014.

House Armed Services Committee, *Ensuring Navy Surface Force Effectiveness with Limited Maintenance Resources*, Washington, D.C.: U.S. Government Printing Office, August 1, 2013.

Malone, Michael, Christopher S. Trost, Courtney Firestone, Tom Mullen, and William K. Krebs, "Maximizing Value Across the Lifecycle of Long-Lived Capital-Intensive Assets," *Naval Engineers Journal*, Vol. 126, No. 3, September 4, 2014, pp. 67–79.

MILPERSMAN—*See* Navy Military Personnel Manual.

NAVADMIN—*See* Navy Administrative Message.

Naval Sea Systems Command, "DDG 51: Fact Sheet," no date-a. As of February 2, 2016:
http://www.navsea.navy.mil/Home/TeamShips/PEOShips/DDG51/FactSheet.aspx

———, *Strategic Business Plan 2013–2018* (second edition), Washington, D.C., no date-b. As of January 18, 2016:
http://www.navsea.navy.mil/Portals/103/Documents/Strategic%20Documents/SBP13-14_Final-2ndEd.pdf

———, *Technical Foundation Paper for DDG-51 Class Type Commander Notionals*, Washington, D.C., February 2012.

———, "DDG 51: Program Summary," Washington, D.C., December 2015. As of February 2, 2016:
http://www.navsea.navy.mil/Home/TeamShips/PEOShips/DDG51.aspx

Naval Sea Systems Command, Deputy Commander, Surface Warfare, *OPNAV Notice 4700 Change Recommendation: DDG-51 Class Depot Availabilities*, 4700 Ser Sea21/126, Washington, D.C., 2012a.

———, *Surface Maintenance Engineering Planning Program DDG-51 Class Technical Foundation Paper*, 4700 Ser 12-045, Washington, D.C., May 23, 2012b.

———, *Technical Foundation Paper (TFP) for DDG-51 Class Revised TYCOM Notional, SEA 05D*, 9830, Ser 05D/327, Washington, D.C., 2012c.

Naval Ships' Technical Manual 050, *Readiness and Care of Inactive Ships*, Washington, D.C.: Department of the Navy, May 1, 2005.

NAVSEA—*See* Naval Sea Systems Command.

NAVSEA21—*See* Naval Sea Systems Command, Deputy Commander, Surface Warfare.

Navy Administrative Message 268/01, "Suspension of All ITEMPO Program Pay and Gate Management," Washington, D.C.: Navy Personnel Command, October 12, 2001. As of January 18, 2016:
http://www.public.navy.mil/bupers-npc/reference/messages/Documents/NAVADMINS/NAV2001/nav01268.txt

Navy Administrative Message 221/14, "Authority to Pay Hardship Duty Pay * Tempo to Compensate for Extended Deployments," Washington, D.C.: Navy Personnel Command, September 26, 2014. As of January 18, 2016:
http://www.public.navy.mil/bupers-npc/reference/messages/Documents/NAVADMINS/NAV2014/NAV14221.txt

Navy Military Personnel Manual 7220-075, *Guidelines for Hardship Duty Pay–Tempo*, Washington, D.C.: Navy Personnel Command, November 18, 2014.

Navy Personnel Research and Development Center, "Projecting the Impact of a Navy-Wide Decrewing Policy in the Navy's Manpower Force Structure: A Detailed Approach," San Diego, Calif., June 1980.

OPNAVINST—*See* Chief of Naval Operations Instruction.

PA Consulting Group, Office of Naval Research, and Navy Major Program Manager for Surface Combatants (PMS400F), *Total Ownership Cost Dynamic Simulation of Surface Ship Lifecycle Sustainment: Final Analysis and Lessons Review*, December 2013.

Seamans, Andrew, and Wm. Brent Boning, *Examining the Rate of Corrosion*, Alexandria, Va.: Center for Naval Analyses, DRM-2012-U-000628-Final, July 2012.

Seamans, Andrew W., Robert P. Chase, and Leopoldo E. Soto Arriagada. "The Effect of CBM on Availability Cost and Length," Alexandria, Va.: Center for Naval Analyses, March 2009.

SECNAVINST—*See* Secretary of the Navy Instruction.

Secretary of the Navy Instruction 5030.8a, *General Guidance for the Classification of Naval Vessels and Battleforce Ship Counting*, Washington, D.C.: Department of the Navy, February 8, 2011.

Soto Arriagada, Leopoldo E., Andrew M. Seamans, and Robert P. Chase, "Budgets and Execution: What Are the Implications of Changes in Ships Maintenance Norms," Alexandria, Va.: Center for Naval Analyses, March 2009.

TFP—*See* Naval Sea Systems Command, Deputy Commander, Surface Warfare, *Technical Foundation Paper (TFP) for DDG-51 Class Revised TYCOM Notional, SEA 05D*, 9830, Ser 05D/327, Washington, D.C., 2012c.

UK Ministry of Defence, "Naval Service Vessel Maintenance," *Value for Money Review*, London, March 2009.

U.S. Code, Title 10, Armed Forces, as amended through January 16, 2014. As of January 19, 2016:
http://www.gpo.gov/fdsys/pkg/USCODE-2013-title10

U.S. Code, Title 10, Armed Forces, Subtitle A, General Military Law, Part I, Organization and General Military Powers, Chapter 6, Combatant Commands, Section 991, Management of Deployments of Members and Measurement and Data Collection of Unit Operating and Personnel Tempo, January 3, 2012. As of January 19, 2016:
https://www.gpo.gov/fdsys/granule/USCODE-2011-title10/
USCODE-2011-title10-subtitleA-partII-chap50-sec991

U.S. Government Accountability Office, *Navy Needs to Assess Risks to Its Strategy to Improve Ship Readiness*, Washington, D.C., September 2012.

U.S. House of Representatives, *National Defense Authorization Act for Fiscal Year 2012 (Parts 1–2)*, Washington, D.C.: U.S. Government Printing Office, Rept. 112-78, 2011.